# The Dog Rules

*(Damn Near Everything!)*

## William J. Thomas

 **Prometheus Books**

59 John Glenn Drive
Amherst, New York 14228-2197

*For Whitney and David, Harry and Tulsa.*

Published 2003 by Prometheus Books

Illustrations: Peter Cook

Inquiries should be addressed to
Prometheus Books, 59 John Glenn Drive, Amherst, New York 14228–2197
VOICE: 716–691–0133, ext. 207; FAX: 716–564–2711
WWW.PROMETHEUSBOOKS.COM

07 06 05 04 03      5 4 3 2

Library of Congress Cataloging-in-Publication Data

Thomas, William J., 1946–
    The dog rules : damn near everything! / William J. Thomas.
        p. cm.
    Includes bibliographical references.
    ISBN 1-59102-130-8 (pbk. : alk. paper)
    I. Title.

PN6231.D68T47 2003
636.7—dc21

        2003012701

Printed in the United States of America on acid-free paper

# *Acknowledgments*

Thank you Roddy Allan for providing the backdrop of
*The Winchester Arms*, "the pub of fine ales and single malts"
in Port Colborne, Ontario.
And to Thies Bogner, one of Canada's premier photographers who
always manages to outdo himself, thank you.
Also appreciated were the efforts of Jake's wranglers—
Nancy and Amanda Giles and Monica Rose.

# Contents

# *The Cross Border Collie Story*

WHAT DO YOU GET WHEN YOU CROSS a border with a Border Collie? A dog that wags its tail internationally? No. Sometimes you get a nice big fat Milkbone.

On a sweltering day in a saner and simpler time, before September 11, 2001, I crossed the United States border at Buffalo with the top of my Miata down and my Border Collie/Australian Shepherd sitting sassily in the passenger seat. He's the smartest, handsomest dog in the world and the problem is, he knows it.

So Jake is calm and cool as we coast down off the Peace Bridge to the immigration booth and I'm sweating and nervous because I'm going down to teach my humor-writing course at the Chautauqua Institute in Western New York and even though everything is on the up and up, I could easily be refused entry on a whim or a misspoken word.

Also, U.S. immigration officers carry guns, which makes a Canadian nervous. As you may know, the deadliest weapon in the Canadian military is our new titanium-tipped, stealth pitchforks. We're not real good at violence.

So hoping for the best but expecting the worst, I remove my sunglasses, turn down the radio, and prepare to be interrogated and then probably pulled over to the office for a car search.

The officer, a bulky, balding man, does not remove his sunglasses and as he wipes the sweat from his sprawling brow, he stares at us without saying a word.

I've got all my answers ready but I do need a question to get started. The awkward silence is broken by my dog, Jake, who looks at the guy and barks. Great! We're going over to head office in a hurry here.

Suddenly, the officer walks quickly toward me and I glance at the black Beretta to make sure it's still in its holster. And faster than I can grasp what he's up to, he reaches in front of me and puts a supersize Milkbone in Jake's mouth. Jake, as surprised as I am, stares at him as drool drops off the end of the treat.

Then, taking off his sunglasses he says to me: "Citizen of what country?"

"Canada," I reply a little too quickly.

And then with this big, broad, beaming smile he asks: "Both Canadian citizens?"

I'm laughing too hard to answer him so Jake responds for me with a thank-you bark, muffled by the Milkbone. At that the guy slaps me on the shoulder and says: "Have a good day." And that's the way the world should work.

That's also what happens when you cross a border with a Border Collie—trust and immediate friendship. It was also a sharp reminder of why that 3,145-mile border between us remained undefended since 1812.

If I spend a day in Toronto, I can rarely make eye contact with anybody. However, if I'm with Jake, I get greetings and smiles and questions like: "Can I pet him?" And that's what pets do for people—they strip away our distrust and self-importance and remind us we have a loftier purpose. Mainly—to look after them.

Dogs and cats know no borders, only love and abiding affection. If we ever figure out how to treat each other the way we treat out pets, this world will actually see the end of this century ... in much better stead than it started.

# CHAPTER ONE

# *From Cat Lover ...*

**I HAVE, FOR HALF MY LIFE, LOVED CATS.**

I never much cared for or paid attention to them until I got married and my wife and I needed pets to protect us from the thought of having kids.

As a boy growing up with two dogs and the next-door neighbor's gorgeous blond collie, I naturally wanted a canine, specifically a boxer.

My wife's family had always kept cats. Lots of them. Of course, she wanted a cat.

So we struck one of those marital compromises, a necessary accommodation of wills that keep a marriage running smoothly. We got two cats and then in no time at all, two more.

Dogless, and living in what was once the servants' residence of a gated mansion in Fonthill, Ontario, I wearily began to watch these little buggers, these arrogant tiny people with four legs and fur.

And gradually, reluctantly, I learned to love cats. I could get down on all fours and outstare them, but I, of course, could never outsmart them. I studied the habits of each of them and figured out ways to mess them up. I never stopped marveling at their ability to relax, to stretch, to nap for days at a time, and turn an open drawer into a pullout couch.

I found joy in their excitement, laughter in their playfulness, and alarm in their anger that would see tails stop wagging and begin snapping back and forth—"Thump! Thump! Thump!"—the final warning that teeth and claws would be deployed if teasing were to continue. I was always amazed that on those occasional special Friday nights when we smoked up, the cats got stoned sooner and seemed to enjoy the Beatles even more than our guests.

When the marriage ended, she got three cats and a dark green Cutlass convertible with a khaki canvas top. I moved to my cottage on Sunset Bay with Malcolm, a buck-toothed, short-hair gray and gangly bundle of love that the other three cats hated. It was a food thing. With the grin of a Cheshire and the swagger of John Wayne, he was the wonderful waif I had for eighteen enchanting years—a tale I recounted in a book entitled *Malcolm and Me: Life In The Litterbox*.

The moment I got Malcolm's ashes from the crematorium, I placed them with great reverence in a highly polished mahogany urn on the mantel of my fireplace. Then I dutifully repeated the solemn oath of all good pet owners: Never again. No more cats. No more pets, period.

And I said so in my weekly syndicated newspaper column, lest some well-intended reader drop a kitten on my doorstep.

The five-page letter from Carole Hallpike who lived on a farm near Caledonia, Ontario, was a most shameless and pathetic attempt to foist an unwanted pet onto a vulnerable and grieving man. There was the physical description of this magnificent, young, haughty male—the white bow-tie front, the white bobby-socks legs that set off a shimmering coat of many colors—gray, brown, beige and rust.

She must have used a cliché dictionary for the last two pages: Clever as a fox, cute as a button, smart as a whip, as lovable and handsome as her husband ... when they'd first married, of course.

It was such a sorrowful letter I'm sure the stamps had been wetted with her very own tears. And I told her as much when I showed up at the farm the day after I got the letter, carrying an empty cardboard box to bring this little orphan home.

After he wedged himself into several tight spots behind rads and between closet doors, I realized I'd been oversold on that "smart as a whip" feature.

Wedgie had more personality than any cat or dog I'd ever known. He was handsome beyond description, with a wicked sense of humor. If I ignored him in the morning when I went out to my detached office to write, I would inevitably hear him scurry up the fence, then bound along the peak of the roof to my window, where he would contort himself upside down, staring at me as he banged on the glass until I'd fetch the stepladder from the toolshed and bring him down. I'm not someone who laughs easily, but this crazy little creature constantly had me in stitches.

He'd walk with me for a half mile along the beach, occasionally running sideways in some sort of mock attack mode, making me laugh out loud. We'd race home, me running along the shoreline and up the breakwall steps, Wedgie sprinting under cottages and over decks until he'd beat me to the kitchen door and be sitting there waiting, preening himself until I got there. He'd sleep under the covers of my bed, and without provocation or warning he'd begin to creep and circle and dart—this large and ghostly lump lurking under the bedspread like Cato in the *Pink Panther* movie. The more I'd laugh, the faster he'd move.

Wedgie's smarts bordered on fiendishness. All cats stalk birds; he hunted them from within their own feeders. The neighbors still have the photos.

When the lake froze over, Wedgie chased the puck around the rink during the neighborhood hockey games. He so charmed his house sitters they'd visit him when I wasn't away, always bringing gifts.

This cat was not a cat, this was a friend from another life come to visit me in a perfectly believable feline disguise. God, how I loved that cat.

The way I cared for him and protected him from Sunset's Lakeshore Road, Wedgie was supposed to be with me for eighteen, maybe twenty, years.

At four years of age, bright and beautiful and a growing legend as the tiny terrorist of Sunset Bay, Wedgie hung himself. I didn't want him out at night, and one Saturday evening in August 1996 he wandered off on a leash I'd failed to secure. He got caught in deep brush near the creek where he used to hunt frogs, and he struggled to get free until he strangled himself to death.

Wedgie was not wearing an expandable collar, the kind that really do save lives. He'd lost three in as many months and he was wearing a fixed leather collar until I could buy more.

When he wandered off at dusk dragging that leash, it could have, and should have, caught on a hundred different and safe things between the patio and the field, but it didn't.

The leash, the very thing I believed would keep him out of harm's way ...

I got up several times in the night but could not see or hear him. I called and searched all the usual hideouts with a flashlight. A neighbor thought he heard some strange and desperate sounds but didn't investigate.

I found him in the morning, still and stiffening up. That was four years ago, and I was forty-nine years old and nothing to that point in my life had left me with such an awful sense of loss.

As the noonday sun sent kids and weekend cottagers splashing into the rippling coolness of Lake Erie, the creekside glen hosted a burial service of three.

Wedgie was wrapped in his favorite blanket with a few of his favorite things encased in a cardboard box two meters below the soft surface of the embankment. Two shovels and a pick were propped up against the tree that will keep him forever shaded.

Over the square of freshly turned ground two guys hugged in grief. One was crying. The other was Tim Laing, a true friend indeed.

Until this point in our relationship, a firm handshake had seemed a sufficient emotional gesture, but Tim has cats and dogs. He *knows*.

Like his calamitous little life, Wedgie's funeral was short and sweet. A few words, some intermittent sobs, his name tag saved for a lasting memento, a dozen rocks to mark the place—it didn't last more than twenty minutes.

Once we were sure there had been no witnesses to the hugging part, we left.

Later that evening at the wake on my patio, I was dumbstruck by my friends Tim and Lee's two Border collies who positioned themselves on either side  of me at the table, nudging my legs and looking up at me in sympathy. Somehow, they knew.

All that day and the next I recited the deceitful dirge of every pet owner in mourning: never again. No more cats. No more pets, period.

We say this for two very good reasons.

First, at the time, we really mean it.

And second, no matter how it all plays out, a true pet lover is one sick puppy. Let's move on.

CHAPTER TWO

# ... *To Dog Fool*

IT'S HARD TO KNOW EXACTLY WHAT CAUSED ME to break my pledge of no more pets—the guilt of my responsibility for Wedgie's death or the ensuing silence that grew louder each passing day in a house where suddenly no creature stirred.

Two things became painfully clear.

Living alone as I have for twenty years is great, as long as there is something small and stubborn around to bring a daily dose of aggravation to your life. (Marriage I found to be an overdose.)

Also, my sworn and sacred self-promises were a lot like that low-fat buttermilk-based salad dressing I buy. Both are good for three months at the most.

At four o'clock on an excessively hot fall day, I was trying to outdistance the midweek exodus from the city of Toronto, trying desperately to get past the bottleneck at Highway 427 South, hoping to sneak through Oakville and make it home to the north shore of Lake Erie by five or shortly after. Rush hour on the 400 series highways surrounding Toronto, constructed for the million citizens of that city and not the six million that now call it home, looks a lot like the road to Baghdad the day the Iraqis fled from Kuwait. You can age, and not very gracefully, in Toronto's weekday gridlock.

Relieved to see Highway 401 fading in my rearview mirror, I came up the ramp to Highway 427 and abruptly turned north. I was supposed to swing left, onto the south entrance, but the steering wheel of my small red Mazda turned right, almost by itself. As Yogi Berra might say: "I didn't even make up my mind until after it was all decided."

I fished out the piece of paper with directions to Pamela Hall's house and put it on the dashboard. Against my better judgment and strongly overruling my free will, the car (with me in it) was headed north to a farm near Palgrave where three dogs lived. One temporarily and, by all accounts, unwelcome as well.

I'd been carrying around this piece of paper, now dog-eared and faded, for three months to the day that Tim and Lee Laing told me I needed to get a dog. They had Border collies. They loved Border collies, and therefore I too needed and would come to love a Border collie. I have to admit, I was still amazed and haunted by that scene of unmistakable solace, Buddy and Maggie, faithful attendants at Wedgie's wake, poking me with pleading eyes: Things would be all right, things would get better, they were saying.

I was once more broadsided by clichés: Clever as a fox, clean as a whistle, fast as lightning and as friendly as a phone-sex provider with her meter running. (Okay, I made that one up.)

For these past three noiseless months, I had thought long and hard about getting a dog and had decided against the whole idea.

I mean, a dog is a lot of work. You don't have to walk a cat and get him out twice a day to make sure he poops. With a cat you just lead him over to the neighbor's flower bed and drop your pants to show him what's expected of him. The covering-it-up-with-dirt response seems to be instinctive.

So why was I driving north to look at some dog who had eaten himself out of two foster homes and whose next stop was the humane society?

Ah yes, to find him a good home. Maybe I'll bring him home and run some want ads in a few local newspapers and save him from that barred concrete cell waiting for him down at the howlers' orphanage.

As I pulled into the driveway of the bright, white hobby farm in the Albion Hills, I could hear the barking of dogs in the backyard.

There I found Pamela Hall playing a game of fetch with three dogs—her two Border collies, Jesse and Shadow, and a larger black-and-white half-breed that stayed pretty much to himself. Only occasionally would this mutt, who was twice the size of the Border collies, give me the courtesy of an occasional look. With a mix of suspicion and curiosity, he'd narrow his eyes then raise a pair of the most beautiful, soft black ears I'd ever seen on a dog. Then he was off chasing the dogs who were chasing the stick.

Like the uncle in the family photo that nobody talks about, this dog just didn't fit in this playful backyard scene.

Pamela would throw the stick high over small scattered pine trees to the back of the lot and all three dogs would tear–ass after it. Her two dogs would take turns winning the race and bringing the gnarled hunk of wood back to her.

For Jake, as I came to know his name, coming third in this tournament of three was getting real old. Bigger, broader and not nearly as fit as the smooth and darting Border collies, Jake was fixated by the stick and could not bring himself to quit playing a game he could not win.

"He prefers women," said Pamela as Jake circled me, more than a little wary. Even with an unblemished record of defeat at this game of stick, his tail remained proudly high, curved so that the white tip brushed his black back as it wagged slowly from side to side.

As someone who has all his life hung out with guys who were not real bright and damn proud of it, I felt a spark of male commonality with this dog who sat panting heavily, trying to catch his breath while the two females crouched, poised, anxiously awaiting the next round.

Jake sniffed my outstretched fingers, dropped his ears and punched my hand upward with his cold, wet nose. I crouched down beside him, and his hazel-brown eyes got bigger as his fear subsided.

I got in one quick pet before the stick went sailing into the air and Jake was off, as usual to a very bad start.

For the first time, Jake barked at the other dogs with a large lead and seemed to run not only faster, but with a fresh purpose in his stride.

Instead of dutifully stopping and trailing the dog with the stick back to Pamela, this time Jake ran right over the first dog and then, blocking the path of the second dog, ripped the stick out of her mouth.

Whirling dervishes until now, the Border collies froze, exchanged stunned looks like this was not right. Somebody broke the rules. Then instantly they were in full pursuit of Jake, who was now growling and gurgling with glee, the prize held high like the Stanley Cup in the traditional grip of a victory lap.

Why, the little rake! In one bold and some would say brutal

maneuver, Jake had taken the despicable and dehumanizing violence out of hockey and ... put it into the game of fetch. A sport, I might add, that could use a little grim mayhem.

Charging, roughing and finally high-sticking, Jake looked like a hockey goon headed for a game misconduct, like Dave Semenko in his rookie season or Marty McSorley in his last game.

And if Pamela Hall and her bonny little Border collies thought they were going to get their stick back anytime soon, there was a second surprise headed their way.

Jake pranced around them like Gretzky on a power play, uttering sharp barks of warning muffled somewhat by the wet piece of wood that kept his mouth ajar and exposed his bright white teeth.

This was not a dog with a stick in his mouth; this was a guy who, at least momentarily, had shrugged off the mantel of loser. This was the English ski jumper Eddie the Eagle bowing at center stage of the Olympic podium as they placed the gold medal around his neck.

"Come!" "Sit!" "Give!"—Jake ignored every command in the obedience manual as he ran rings around everything and everybody. He bobbed, he weaved, he teased and he tormented, but he did not—and would not—give up that damn stick.

I tried as hard as I could to look as concerned and disapproving as all three of the females, but deep down I was cheering and silently yelling: "Atta boy!"

By my watch, Jake's fifteen minutes of fame lasted just about fifteen minutes. Then we all went into the house to do the deal.

In the finished basement of the Hall home, Pamela went over what amounted to adoption forms as Jake lay by the door, paws

crossed in front of him, the stick by his side. He was an impressive animal, a shiny black body and face interrupted only by a snow-white chest and front legs matched by two white patches at the back of his neck and the tip of his tail. He was slightly overweight with a somewhat awkward aura. What I discerned in him to be a flash of affection was abruptly stopped by a lingering leeriness of his surroundings. And me.

Staring at each other from across the room as Pamela went over my suitability as an owner and Jake's checkered history, it seemed to me at that moment we both needed the same thing. And neither of us understood what it was.

As Pamela detailed the last three foster homes in which Jake had lived, he stared at me with an equal measure of distrust and bore-dom, as though I would be his next victim. He figured it was the same story with new surroundings—but at least there'd probably be a pretty good car ride involved.

For the first part of his life, Jake had been the pet of a young unmarried couple. He was purchased from a farm near London, Ontario, where he was the runt of the litter and the last to be taken. His parents were working dogs. The couple split and, according to Jake's file, neither could take him into their new accommodations. (Four words for people abandoning four-footed family members to the Draconian "No Pets Allowed" rule: FIND ANOTHER %@#*ING APARTMENT.)

Born January 1, 1992, Jake had spent three years with these peo-ple before being put up for adoption.

Enter the North American Border Collie Rescue Network Inc. This is a lifeline for unwanted Border collies, a benevolent foundation with an office and hot line in Ithaca, New York, which keeps track of all deserted dogs of this breed in a comprehensive database. (You gotta love it. In the United States, they may never know who shot President Kennedy or what caused TWA Flight 800 to plunge into the Atlantic off Long Island; and in Canada, we may never know who's responsible for the APEC riot, the tainted-blood scandal or the Airbus fiasco, but dammit, we know where every single Border collie is on this continent at any given time.)

Out the door one day and onto the rescue foundation's computer the next, Jake had spent months bouncing around the homes of var-ious volunteers. In Collingwood, Jake was just too much for a woman who had taken in three unwanted collies. He was upsetting the routines of her own five.

In Brampton, nobody knows for sure what Jake did, but the woman not only kicked him out, she canceled her membership to the association the very same day.

Jake then lived in Toronto with an evaluator long enough to have his history and behavior recorded, and then he was passed along to Pamela, who'd given him a home for the last three weeks.

She told me she really liked Jake but could not keep him.

"Likes to chase cats," read the report. "Behavior around children unknown."

Jake was constantly chasing her two cats, something he enjoyed even more than fetch. Although recently he'd lost a lot of his enthusiasm for feline fetch when one cat purposely took off in the backyard, enticing Jake to chase her, and then whacked him numerous times once he got close enough. (One of the reasons I believe cats should be offered teaching positions at all dog obedience schools.)

"Very friendly, slightly food aggressive. He's presently on a diet which, with increased exercise, has made him very hungry, so he doesn't like to be bothered when he's eating." Who among us does? I thought.

"Food aggressive" was the term Pamela kept coming back to as we went over Jake's file. That was the nice way to put it. I now understand why Jake was so wide and you could see the ribs of Pamela's dogs. Her Border collies were being fed twice a day, but they hadn't eaten in weeks.

Jake occasionally sighed and shifted as Pamela read his rap sheet:

"Spayed with all his shots, very friendly except when interrupted while eating, doesn't work or herd but could learn, likes to play fetch but needs a lot of work giving the ball back."

Tell me about it, I thought. Barring some divine dog intervention, the stick that lay beside his paws was on its way to a place called Wainfleet.

"Jake's a very happy dog who bonds well," claimed the report, "if he gets a lot of attention, exercise and food. He takes grooming very well, likes to sit next to you and will sometimes even cuddle."

Pamela told me how she'd once taken Jake to a flyball tournament, a relay game where a team of four dogs race up and down designated lanes vying for the fastest time in retrieving a tennis ball. It's a hugely popular international dog sport dominated by speedy Border collies.

Pamela's two dogs played the game, and she assumed Jake would love it. And he did. As the other dogs frantically raced up and down the track, Jake ran full tilt ... for the bleachers. There he sat, eager and watchful, with the other spectators who had gathered to enjoy the game. I imagine he looked rather regal sitting with the humans, clearly superior to those "carny" dogs in the fray down on the field.

Apparently he wanted nothing to do with the sport itself. No. I'm guessing that what Jake really wanted, besides a front-row seat, was a guy coming around selling chili dogs and cold beer. As a fan of the game, Jake did everything but start the wave.

By obligation Pamela read me "The Posting," a warning to would-be owners that pure Border collies are essentially smarter than your average local elected official and slippier than Cool Hand Luke. They can jump six-foot fences, dig under walls, open doors and gates and, in an emergency, hot-wire any automobile manufactured before 1998. A Border collie is most content when he has a rewarding job to do. As a freelance writer for the last fifteen years, I can absolutely relate to that.

Then I became the focus of this interrogation. What kind of an owner would I make? Single, lives alone, works at home all day, has a house on a large lot (a winterized cottage, actually, with miles of beach and nearby fields and trails). Not had a dog since grade school but has owned two cats, both of which died. How? Naturally. Okay, naturally and accidentally.

Was I willing to give this dog the attention, exercise and mental stimulation required to keep him happy? Short of getting him a library card, the answer was yes, I said, crossing my fingers behind my back. Or find somebody else who will. The want ad option was still open.

I signed the agreement and was so impressed with this organization, which is approximately ten times as efficient as any level of government we now know, I wrote a check for more than double the transfer fee of $75. Jake seemed completely unimpressed that this was the second time he'd been purchased in less than four years, or that just as a racehorse who keeps winning, his claiming price was increasing dramatically. Like a clever auto dealer, I decided I was not buying a "used dog" but more a dog who was "pre-owned."

The intense suspicion returned to his alert black face as I attached him to a short leash and picked up a plastic bag of his belongings. A spare leather collar, some old tags and a half-chewed ball—for a dog that had been on a two-year tour of the province, Jake sure traveled light. I put the four sides of his metal cage in the back seat and Jake in the front. For better or for worse, for a day or for a week at the most—I had a dog.

It was probably my imagination but as we pulled out of the circular driveway, I thought I heard Jesse and Shadow applauding. Nice dogs, but they never got their stick back.

Whatever goodwill that passed between Jake and me in Pamela Hall's backyard and basement quickly dissipated on the long ride south to Wainfleet. It was a hot and humid drive, and Jake slobbered all over everything—the seats, my shoulder bag, the dashboard, my right arm.

This was new and not a little offensive to me, an animal panting and goobering saliva in all directions. Cats don't pant and drool. A cat would have found the coolest spot in the car, probably under the passenger seat, and settled in for a long sleep.

But this, this sad-looking beast acted as if he was suffering from heat exhaustion. Leaning far to my right, I rolled down the passenger-side window, and instead of sticking his head out like every other dog in the world, Jake sat backward so that now he was drooling on the headrest.

It was a very unpleasant three-hour trip for both of us. Jake stopped panting just long enough to look at me with grave mistrust, like this was the last ride, the one that ends at the humane society, or yet another trip to a halfway house inhabited by other people's pets.

Every now and then I found myself staring at him and thinking this dog has led a life of five years that I know almost nothing about. This dog could have killed somebody or saved a family from a fire, and I'd never know anything about it.

By the time we got to Sunset Bay in Wainfleet, both of us were very damp and completely convinced this had been a big mistake in both our lives. Oh yeah, Jake would be starring in a whole series of want ads real soon.

Jake seemed uninterested in his new surroundings; he just sat by the patio door at the kitchen and waited for the next move.

Too tired to assemble his cage, I spread an old blanket near the couch in the living room, brought him in and told him to lie on it. He did.

I tried to play with him, rough him up a little and roll a ball over the carpet into the TV room. But all he did was sit at attention and stare at me. Whatever he was anticipating, it wasn't fun. When I got tired of playing fetch with myself, I went to bed to read. Every time I checked on him, he was exactly the same, sitting, waiting, staring at me, waiting for something to happen. We would eye each other several times in the early evening, embarrassed, but tolerating an uneasy arrangement neither of us obviously wanted.

Since I had no intention of bringing home a dog that day, I had nothing for him. I filled a margarine tub with tap water, and onto that day's newspaper I chopped up some leftover steak and potatoes.

As I prepared his meal, I could hear him licking his lips from the other room, from the blanket that had become his home base. He was starved and salivating yet when I called him to dinner he still sat at attention on that blanket. He wanted so badly to do the right thing ... there must have been an "un-stay" order, I thought, that nobody thought to tell me about. So I gently led him by the collar and led him to the food and water in the kitchen beside the refrigerator. Once there, he sniffed the food, gave me a very grateful look and launched into it, eating slowly as if savoring every bite.

Watching him eat with such purpose and pleasure, I now knew why his parents were working dogs. Even as a puppy Jake must have racked up a helluva grocery bill.

He washed it all down by emptying the entire tub of cold water and returned to his security station, the blanket.

Mutual indifference reigned, and I went to bed, shaking my head in disbelief that I could be so insecure as to bring home the first domesticated animal I looked at since losing my cat. Things would be fine, I told myself. I'll find him a good home—on one of the hundreds of nearby farms. And then I'd get myself another cat.

They're a very different species, I thought as I got ready for bed.

Having a cat is like having a pet.

Having a dog is like having a really stupid brother who needs a lot of looking after.

# The Dog Rules

As They Apply to Cohabitation

A social system without rules is like a veterinary clinic
without cages—only anarchy can reign.

### Rule 1
The dog owner/guardian/home owner of said dog makes the
rules. (And for the record, the dog never "said" nothin'. Yet.)

### Rule 2
The dog, as pet/ward/resident of the owner/guardian/home
owner obeys the rules. And the owner is always right. (Which
is not to say that the dog is always wrong.)

### Rule 3
And even if the owner is occasionally wrong, those rare, albeit
"wrong rules" still require complete compliance by the dog.

### Rule 4
Okay, let's say you say "Speak" and he barks
instead—that's close enough.

### Rule 5
Okay, to be perfectly fair, if the owner's rules are wrong
or vague, say, half the time, the dog only has to listen
and obey half the time.

### RULE 6

But if you say "Sit," and the dog noses the top off his earthen-ware cookie crock and helps himself to two extra-large Milk-Bones then runs into the den and eats them, leaving crumbs on the couch—well, that's just … close enough.

### RULE 7

The dog does not make the rules. That would lead to household chaos and the breakdown of western civilization … such as it is.

### RULE 8

Okay, the dog could make up some house rules, but the owner is in no way obligated to follow them unless it's the look of distress on the dog's face at 3:00 a.m. that says: "I GOTTA GO! NOW!" This rule cannot be ignored.

### RULE 9

Okay, so a lot of the dog rules, say, the majority, can be made up by the dog, and if the owner knows what's good for him, he will obey these rules, but if he disobeys a rule and the dog doesn't find out about it, the owner is under no obligation to confess.

### RULE 10

No, there will be no shame-based justice. We will not wear signs in public that read: "I'm a bad dog owner." No way.

# CHAPTER THREE

# *The Beginning of a Beautiful Friendship*

I AWOKE AT EIGHT O'CLOCK THE NEXT MORNING, pulling the bedroom drapes back to reveal an already hot sun rising over a calm and cool lake. With no bathers on the beach yet, I pulled on a pair of shorts and grabbed a towel for a quick swim before the kettle boiled for tea.

As I passed from the bedroom to the kitchen I noticed first the old blanket, out of place and in a heap on the living room floor. Then something or somebody stirred slightly on the couch above it.

I swear: I'd forgotten I had a dog!

I stopped, staring in groggy disbelief at the scene in front of me. There, surrounded by the dark pine walls and framed by the cathedral beamed ceiling was one of the strangest sights I've ever seen.

Jake, yeah that's his name, was lying on my brown corduroy couch, on his back with all four paws stretched straight up into the air. His front paws were set together like those of a praying mantis, his back legs splayed apart. And he had what can only be described as a demonic smile on his face, the kind you see in cartoons where dogs talk in dialogue bubbles. His nose twitched and his lips moved slightly up and down and his teeth verily gleamed in a village-idiot grin.

Past the couch in front of the fireplace on the thick white, wool carpet—the one I'd hauled on my back from Marrakesh to Amsterdam for a flight back home to Canada in the late '70s after a year of backpacking around Europe and North Africa—there, on that beautiful plush carpet that I'd taken a whole day to haggle over with a Berber trader and his gallon of sweet mint tea, right there in the middle of my favorite carpet was a large mound of ... how can I put this ... okay, dog shit.

And this ... this mutt lay there on his back, wagging his tail and smiling as if to say: "This is great, Bill! Steak and potatoes for supper, I get to sleep on the couch, and the best is, I don't even have to go outside to go to the bathroom. What a deal!!!"

I can't tell you everything that transpired next amid the screams, the growls, the threats and the use of a common bathroom towel as an effective offensive weapon.

Once I'd calmed down, once I'd cleaned up the mess, once I'd decided not to kill him right there on the spot—I decided to keep Jake.

First of all, in this pathetically politically correct world in which people wouldn't say the "s---" word even if they had a mouth full of it, I had to admit this was a pretty bold statement by either man or beast.

Second, it was funny. I mean, once the color returned to my face and I recalled his toothy, satanic grin, I had to go over to my detached office, close the door and laugh for a very long time. But not so loud that he could hear me, eh. This kind of canine toilet humor should definitely not be encouraged.

That was five years ago, the day Jake and I enrolled in Intensive

Dog Training 101. And I have to admit, it's been slow and sometimes painful, tedious and occasionally rewarding. What can I say? I'm a slow learner.

# CHAPTER FOUR

# Settling Right In

THAT DAY, THE FIRST FULL DAY I HAD HIM, Jake figured everything out in less than an hour. Walking around inspecting the place carefully, his expression alternating from cautious curiosity to relieved familiarity, he settled right in. In a magic moment, he even bonded with me by climbing onto my lap in the office. The rest is simply an unfolding history of two not-so-young-anymore guys who will be separated only by death.

First he ripped up and down the sandy shore, chasing the white curl of the waves and working himself into a heavy pant. Then he tried to drink Lake Erie, and when he realized he couldn't empty it, he gagged and coughed a couple of times and decided to swim in it instead.

I had to coax him in at first and by walking slowly backward, I got him to go farther and deeper on each turn before he circled and swam back to shore. He kept to himself while swimming until he realized I wasn't going to splash him or dunk him, and then he got braver and came closer, and soon we were both way out over our heads. I started back to shore, assuming he'd follow. He spotted a boat, a large laker a mile off shore, and by the time I got to where I could stand up, Jake was well on his way to herding it in. He was swimming directly toward Cleveland, Ohio, before I convinced him to turn and come back.

A couple of shakes and three rolls on the grass, and he was practically dry. He's a very fast-drying dog.

After a breakfast of yet more chopped-up steak and potatoes, Jake took a complete tour of the property. He barked at a few neighbors, marked nearly every one of the seventeen silver maples on the lot and nosed his way into my outside office. After taking him up the long gravel driveway and explaining very loudly, "Road! Bad road! Bad road!," he bolted back to the house and has never stepped foot on Lakeshore Road since.

The experiment to tie him to a tree with a long rope lasted about three noisy minutes. He barked incessantly. I let him go, and he stationed himself exactly where the rope would have placed him: A poised and permanent picture of self-restraint.

Later I went to work at my desk, and Jake burst through the door, jumped up, with his paws on the arm of my wooden swivel chair, and we had another one of those magical two-guy moments. The more I flattered him, the more happy and affectionate he got, eventually working his way onto my lap. With all sixty pounds of him kissing and whining with joy, bonding, I learned, was not something to be taken lightly.

Yes, I said, you stop using my Moroccan rug for a latrine and this living arrangement could well work out.

As I quickly learned, any kind word or soft voice sent Jake's tail spinning and his teeth bursting out in a smile. And that was it. He bounded out of the office and situated himself on a patch of grass from which he could see my office door, the kitchen door and the road. That would become his vantage point, the security hub from which he could observe and control all the spokes of household activity.

Instantly he established the routine: Any and all passersby would be barked at for safety reasons only, any vehicles coming up the driveway would be inspected and their occupants sniffed, all squirrels would be treed, other dogs would be approached and warned off but not attacked and my every move would be carefully viewed with erect ears, inquisitive eyes and a slight tilt of the head. "What are we gonna do next, Bill?"

I swear, within an hour, this dog selected the central point of the property as his post and has run the place ever since.

"I know my job, and I won't let you down," he was saying as I passed from office to kitchen and back again and saw his tail move slowly from side to side as he sat at attention scanning the perimeter of the property where trouble might intrude on us.

Some days I can pass him twenty times, and it's always the same nod, smile and tail wag that says: "All's quiet on the front lines, you just worry about your work." The dog watch (see The Dog Dictionary) security system was now officially up and operating.

Like the fastest of friends, but knowing very little about each other, Jake and I set off on the adventure of cohabitation. He knew his place almost immediately, and now, five years later, I'm starting to understand mine.

Late in the afternoon, I took Jake into town to buy him some proper eating and drinking gear, a collar, a leash and a lot of food.

You realize the real difference between a cat and dog when the bonus you get after carrying the bags of food out to the car is a sore back. Wouldn't home-based grain elevators make a lot more sense for dogs?

Jake got to see some of Wainfleet and the city of Port Colborne from the passenger seat of the car. All backwards, of course.

And I pointed out all the highlights along the way, talking to him as if it was the most natural thing in the world.

How the hell did this happen?

I mean, even with my cats I was loony, but not completely certifiable.

I would talk to my cats, of course, but in a clear and controlled conversational voice. Cats don't like to be talked down to.

But this dog! At any given time somebody eavesdropping on my house or office would hear an otherwise normal human being, in a stupid baby-talk tone, saying over and over again: "Who doesn't love this bobo?" "Who doesn't love this bobo?"

Dog people take pride in embarrassing themselves. Put twenty of them on a bus with Jack Nicholson in the driver's seat and you could call the whole sick syndrome *One Flew over the Doghouse*.

Yet the question remains and it is a valid question: "Who doesn't love this bobo?"

# CHAPTER FIVE

# Health Care—No, Not Yours, the Dog's!

IN THE FIRST MONTH I HAD JAKE, I learned the hard way why he'd been in and out of four homes—he's a dog that's big on accidents but short on health insurance.

Early on, Jake proved to be a world-class klutz. Including his introductory checkup and shots, this dog cost me $400 in the first month, roughly the same amount as my heat, hydro, water and telephone charges during that period of time. (Which I might add, he enjoys but does not pay a portion of.)

On a long walk through dense bush across from my house, Jake seldom stayed on the trail. He was off leaping over dead logs and dodging low limbs in his quest to tree every squirrel in Wainfleet Township, simultaneously.

Suddenly there rose a sharp yelp from somewhere ahead, and from down the embankment a whimpering dog came limping up the trail, stopping, and then flopping to the ground. I couldn't see anything bleeding or broken, but he would not walk on his own. So I picked him up and carried him the half mile out of the thicket to the road. Panicked, I broke into a labored run to get him out of there and into the car for the mad dash to the vet's office.

By the time we reached my driveway, he had stopped whining in pain, though I was damn near dead.

I put him down and a funny thing happened—he took off after a squirrel.

That was one helluva performance to win a free ride home. Actually he was slower than usual, so something was wrong, but nothing that would warrant the theatrics out on the trail.

Although I dodged a vet bill on that episode, I wouldn't get so lucky three nights later when I spent the evening with one hand down Jake's throat, the other shining a flashlight into his mouth.

I was convinced that whatever it was that had caused him to throw up in every single room in the house was lodged in his gullet, and I was going to get it out myself since my local veterinary clinic had closed hours before.

It could have been the splinter of a stick, the bone of a dead carp, a pinecone or a plastic fork, all of which Jake seemed to have an appetite for and which were readily available on the beach.

Repeatedly I probed and he gagged, but I could feel nothing that wasn't supposed to be there.

My friend and neighbor John Grant, a much more experienced dog owner, rushed over, and he went even deeper down Jake's throat than I had. Nothing. He too was amazed at how trusting this dog was and how he allowed a strange hand to grope his neck from the inside.

The dry heaves and hacking continued until 1:30 a.m. when I bundled him into the car and took off to the all-night emergency veterinary clinic in St. Catharines, thirty miles away.

It was a quiet night as crisis clinics go, and the young vet, a recent college graduate, I guessed, had the time to see Jake immediately, and she was a true dog lover as well. Jake took to her immediately, and she fawned over him, gently searching for the problem as I had, but referring to it as his "booboo."

Unsuccessful in her perfunctory exploration, she led Jake into the back for a more complex examination.

Fifteen minutes later she emerged with good news and a bill for $106.88.

As she knelt beside him and fluffed up his ears she delivered her prognosis.

It seemed that the two hands and one flashlight down the poor little bugger's throat may have in fact dislodged the obstruction, but he had continued to gag from the raking of tissue and the resulting inflammation. Or, as she put it in a playful, childlike voice: "Daddy fixed Jake's booboo! Didn't he?"

Now, I know she's a sweet girl and probably a good veterinarian, and her love for her patients is wonderfully obvious, but I'm sorry—for $106.88 I don't want to hear "Daddy" and "booboo."

For that price I want proven scientific terminology that I can barely pronounce and couldn't even begin to understand. For a hundred bucks I want to hear about a disengaged dysphagia or crisis gastritis or a bunch of malfunctioning masticators.

For seven dollars a minute, I expect a valued medical opinion on unleaded gastroenteritis.

Tempted to make the check out to "Jake's Doggie-Woggie Doctor," I didn't.

I spent what was left of the night propped up in bed with Jake at my side licking cough drops off my fingers. Every hour or so, I spread a molasses-flavored lubricant on the roof of his mouth, which soothed the scratches in his throat.

Last week we bonded; this week we pretty much stuck together.

The lubricant was also a laxative, so as he dozed off between the sweet treatments, I began to believe that this crisis, like whatever object might right now be bouncing off the walls of his digestive system, would pass. (I'd just as soon not be there when it happens.)

In a few days Jake was sufficiently recovered to make the hour walk into Port Colborne with me. Except the word "walk" appears not to be in Jake's twelve-command repertoire.

The four-mile beach walk follows the north shore of Lake Erie from Sunset Bay, past Camelot Bay, and around Sugarloaf Point into town. It's all sand except for a rocky promontory that surrounds the Rathfon Inn.

This is, of course, the spot where Jake chose to take off like a bat out of hell after something—something unseen, a noise or the shadow of a cloud perhaps. He spun out on the first craggy curve, jamming a front paw in a crevice and taking a header over the rocks worthy of Don Cherry's *Rock'em Sock'em* hockey video.

He skinned his chin and right leg badly, cut the inside of his mouth and the right side of his face, and separated two toes on his left front paw.

For the second time in two weeks I had to lift this big lug and carry him home. I'm sure the neighbors were beginning to believe I had adopted the laziest dog in the world.

The good news: this one-dog pileup occurred during normal working hours, so I was able to take Jake to my vet in Port Colborne, Dr. David Thorne, the Marcus Welby of the leash brigade.

David determined nothing was broken and patched him up on the spot.

Anxious to get Jake home, I left without the bill, and as I drove back to Wainfleet with the accidental patient sitting bandaged in the seat next to me, backward, of course, I began to total things up. Let's see: two superficial abrasions, two cuts, one external, one internal, and a badly splayed paw. Okay, so if one booboo costs $106.88, five booboos during non-emergency hours must be ...

We keep hearing that our Canadian health insurance program is going to the dogs, and as the proud owner of a dog who's averaging three catastrophes a month, I for one can't wait. With a government-funded health-care card hanging from his collar, from now on I can send Jake to the vet's in a cab and stay home and get some rest.

Have you noticed that people scrutinize every tax dollar spent on health care for themselves, but hardly notice the dollar figure on the vet check in the rush to get the little creature home?

It's not that our pets are more deserving than us or more compassionate than us or more loving than us, it's just that ... who the hell am I kidding? Of course they are. That's why we're absolutely honored to keep coming second in this man/dog matchup of two.

It's no secret that in Canada we have established a two-tier health-care system. First comes the medical needs of your dog and, if time and money permits, you're next in line.

43

# The Dog Rules

## As They Apply to the Family Car

More and more I see cars and trucks traveling on public
thoroughfares with what appears to be a dog in the
driver's seat. I am utterly opposed to dogs driving cars
because there's no doubt in my mind that they'd always
be trying to bump off cats and making it
look like an accident. Therefore …

### RULE 1

Dogs are not allowed to travel in
motorized vehicles. Period.

### RULE 2

Okay, a seeing-eye dog is allowed to travel in a vehicle
with its owner provided that the owner is not
the operator of said vehicle.

### RULE 3

Okay, in a medical emergency, a dog may be taken
directly to and from a veterinary clinic in a moving vehicle.

### RULE 4

Okay, if the family's out for a Sunday drive,
on Sundays only, the dog can go along
provided he sits in the back seat.

### RULE 5

Okay, the dog can sit in the front seat on the passenger
side provided there's no passenger in that seat.
And if it's really hot, he can roll the window down,
but not all the way.

### RULE 6

Okay, the dog can hang out the side window like a
misplaced hood ornament going 60 mph with tears and
saliva spraying the passengers in the back seat.

### RULE 7

The dog is never, repeat, *never* allowed to operate
a moving vehicle. That would be insane.

### RULE 8

Okay, the dog can sit on the driver's lap with his paws
on the steering wheel, but he's not allowed
to put his foot on the accelerator.

### RULE 9

Okay, the dog can steer the car and work the pedals, but
he's not allowed to drive at night or take the car out alone.

### RULE 10

Okay, the dog can drive at night by himself, but not if he's
been drinking. No way. That's against the law.

# CHAPTER SIX

# *Doghouses—Yours, Mine and the Rich*

RIGHT NOW, SOMEWHERE IN THIS COUNTRY, some guy, well intended, mind you, but nonetheless not real bright, is building a doghouse. And the dog is standing nearby, watching, panting on the outside and laughing on the inside because he knows he's not going to live in this postgraduate shop project for more than a week. The dog knows that in no time at all this doghouse is going to be sitting at the end of the driveway with a "Hardly Used/Best Offer" sign on its roof, the star of the early-summer lawn sale season.

The woman shaking her head at the kitchen window is also not being deceived. She knows that love of a pet is stronger than four-inch nails, and pretty soon both she and her man will be unsuccessfully fighting that mutt for the bedcovers.

Her most immediate concern is that there's not nearly enough headroom for the soon-to-be occupant of the doghouse ... her husband.

Nonetheless, the man plunges ahead because he's a guy, and a guy with a concept and the tools to build it is like the "Energizer Ass," stubborn but endlessly enthusiastic. And kind of cute, thinks the woman, with his bib overalls and the pencil stuck over his ear.

The dog, of course, is just mocking him. "Woof!" (Pause) "Woof! Woof!" barks the dog.

"Right," says the guy. "A roof!" And he's off to the hardware store for a no-stick piece of plastic that snow and ice will slide off of easily.

The dog stands and stretches, and the guy's right there with the measuring tape.

"Okay, wide enough for Sparky to turn around in, not so long that it'll lose his body heat in winter." Then he incorporates an entrance separate from the sleeping area, which will be positioned away from the prevailing wind.

The dog sits and out comes the tape again because the entrance has to be four inches higher than the dog in the resting-bum position. That's because when he howls in the middle of the night, waking up the neighbors and successfully wangling his way indoors, you don't want him to hit his head on the "woof."

And it goes on like that—portability for shade in the summer and sun in the winter, insulation of covered Styrofoam sheets, hinges on the top for easy cleaning, shavings or straw over a carpeted floor, elevation on blocks for dryness and a detachable flap for winter-weather protection—and presto! Whaddyagot? A broadloomed shed for garden tools and bags of peat moss that fortunately fits astride a large wheelbarrow so you can relocate it to the back of the property where it will stay until it serves as the basis for the July 1 bonfire.

Because by the time this guy puts the second and final coat of stain on the doghouse, Sparky is sitting on the couch in the TV room watching Bob Villa grouting the base of a low-frequency flush toilet, which the dog is definitely interested in, especially in the dead of winter when the walkway outside ices over.

So, basically there are three types of doghouses: the homemade kind, the prefab model and the ones I'm looking at in a swank magazine article, which are small palaces built strictly for the well-heeled pedigree.

Some doghouses are so exclusive and expensive, they've spawned a new design science known as … wait for it … *barkitecture*. Honest.

The first full-color photograph shows Coco the Westie sitting just inside the foyer of her Roman temple, which is classically appointed with marbleized columns on either side of the entrance, terra-cotta ornamental finishes and a gold dome at the top. All is quite symmetrical and decidedly Etruscan in the early Romanesque tradition. This so-called doghouse has "rich bitch" written all over it.

Coco the Westie looks like a furry Caligula only confused, like a Roman dog who doesn't know whether to poop or stick a feather down her throat.

The second photo shows a Florida art deco canine home, an airy Caribbean-style cottage with a clear and corrugated fiberglass roof that allows a gentle sun to dabble down into the residence while filtering out the harmful rays.

I'm guessing this home of golden retrievers Hobie and Ryely, who appear in the photo (looking bored), is located in Palm Beach, Florida, where a palm tree is mortified by the yellow stains near its base.

This dark blue and pale salmon pavilion has—I am not making this up—pink flamingos on the front lawn. And yes, as a matter of fact, they do look just as tacky in front of doghouses as on your neighbor's lawn.

And such immaculate dog digs! How could the house of two male dogs be kept in such spotless condition? I wondered. Then it

occurred to me: of course, Hobie and Ryely probably have a French poodle come in once a week to clean. It's the kind of house that if you're an average dog in the same neighborhood, you'll kill to get invited to Hobie and Ryely's for Friday-night card games.

I'm sure with *barkitecture* just taking off, the next generation of doghouses will be designed with individual breeds in mind. The house of a boxer will come with a weight room and speed bag; Irish setters will have Guinness on tap in the Pooch Pub; and Border collies will have a small tract of lush grassland out back where they'll keep their own herd of sheep.

It's funny how some people simply have more money than brains. But their dogs don't seem to mind a bit.

As I said, there are three types of doghouses: The kind you build and convert to a toolshed; the kind you buy and soon sell; and the kind you design and construct like the Taj Mahal, which ultimately earns you a housewarming E-mail from Bill Gates's cyber-pooch.

# The Dog Rules

There is no more personal or practical gift than building
a doghouse for your special friend. Even buying one,
bringing it home and making it a natural
and necessary extension of your own home
is a day you won't soon forget.

### RULE 1

The dog never goes in the main house. The dog stays outside
in the specially designed wooden compartment named, for a
very good reason, the *doghouse*. Remember that "practical
gift" and "natural extension" stuff? Are you retaining nothing
as you read this book? Man …

### RULE 2

Okay, the dog can enter the main house, but only if his own
house is being renovated or repaired.

### RULE 3

Okay, the dog can enter the house for short visits or if it's his
birthday. Or if it's really hot outside, or really cold. Or dark.

### RULE 4

Okay, the dog can stay in the house on a permanent basis
provided his doghouse can be sold in a lawn sale
to a rookie dog owner who doesn't yet know the rules.

### RULE 5
Inside the house, the dog is not allowed to run free and is confined to a comfortable but secure metal cage.

### RULE 6
Okay, so that he doesn't look so much like a prisoner, the cage door can be left open.

### RULE 7
To prevent further back problems caused by you crawling into the cage and coaxing the dog to join you, the dog can sleep on a blanket beside the cage.

### RULE 8
Okay, along with the doghouse the cage becomes part of a "two-for-one" deal in the lawn sale, and the dog can go wherever the hell he pleases in the house.

### RULE 9
Okay, the dog rules the property, and he can run around the house barking his fool head off like he owns the place, but his name should never actually appear on the deed.

### RULE 10
Okay, the dog's name can appear on the deed as the official owner as long as he names you beneficiary in his will. That's the least he could do.

# CHAPTER SEVEN

# *Pet-Proofing Your Home*

OKAY. SO YOU REFUSED TO LISTEN TO THE PEOPLE at the humane society a few months ago when they told you that giving the kids a pet for Christmas was a real bad idea.

You also didn't heed the warning of the vet columnist who said that bringing a pet into the family during the hectic holiday season is like your child being born onstage at a New Year's Eve rock concert.

You now have this little creature walking around the place, bumping into door frames, in a permanent state of shock because he cannot fathom the fall from being the most cuddled, coddled, cutest little bundle of joy Santa ever delivered, to his present household status, exactly equivalent to the recycling box in the corner of the kitchen.

The children, who cried when they had to go to bed without the pet on that first night, have now forgotten his name.

At present, you are doing all the pet chores you outlined in your introductory lecture to the kids, and you're so stressed out that the last little yellow puddle you cleaned up actually belonged to you.

I am here to help. I am a self-taught pet expert who believes in the principle of "tough pet love."

I've learned an awful lot by trial and error with my dog Jake, and unless you, as the head of the household, train your animal to follow a strict routine of rules and regulations that you forcefully introduce and adhere to in the strictest manner possible … Just a second.

Sorry, that was the taxi driver at the door. Jake's taking his cousin Wilson to Winchester's Pub for a few beers. It's Wilson's birthday.

… Where was I? Oh yeah, simply put, you gotta teach the little bugger who's boss.

Properly cared for, your pet will bring you and your family a lifetime of love and happiness, not to mention a mountain of fecal waste material, which you alone will have to deal with because the kids find it to be "omigod gross."

Sure, pets give you more attention than your spouse and less lip than your kids, but they're still animals who are toilet untrainable.

Here then are a dozen steps to success and happiness from my New Pet Tough Love Program:

1. When you first bring a pet home, keep him inside the pet carrier until he adjusts to his surroundings. This will calm him as well as teach him that the pet carrier, under certain circumstances, can serve as a paddy wagon.

2. Cats are as perceptive as some people and more sensitive than most. Immediately remove all photographs of former pets. If you have the ashes of previous pet loved ones on display, do not point to them in a "behave or else" training technique.

3. Much like your mother-in-law, pets are curious creatures who will examine every nook and cranny of your house. They

will not, however, point out cobwebs or run their fingers along the tops of the door sills. Give them full rein of the house. Trust me; with them, the liquor cabinet is safe.

4. If there is any question in your mind whether you should spay or neuter your new pet, simply stand back and look at your children, then look at your husband. How fast can you say the word "vasectomy?"

5. Young puppies and kittens love to eat shoes and slippers. However, some pet specialty shops now carry booties for pets. I advise you to buy your pets their own footwear. They'll still eat your slippers, of course, but now when you catch them doing it, you can sit down right in front of them and eat theirs!

6. Like some non-adventurous humans, pets are very territorial. If you can cordon off the area around the couch in the TV room with traffic cones and razor wire, restricting the movement of both the pet and your partner, your vacuuming days are over.

7. Roughhousing—like wrestling on the rug and rubbing his face really hard—could prove fatal, especially if your new pet is a goldfish.

8. Pet-proofing your home is absolutely vital. Unplug or cover electrical cords for puppies and kittens, keep fishbowls well away from cats and never let the children involve the budgie in their badminton tournament.

9. Never let your pet Vietnamese pig near a BBQ where men have been drinking heavily and have already burned the burgers.

10. Never say a word while watching your husband admire

his hair in the bathroom mirror after unknowingly using Rusty's flea and tick shampoo.

11. Never let your cat or dog sign anything presented to them by a door-to-door salesman.

12. Learn to live with the "in/out, in/out, what's it going to be? in/out," routine and don't let the little grifter talk you into—they're on the market, honest—the easy-to-install wireless Doggie Doorbell.

Certainly there is wiggle room—sorry, certainly there is some flexibility in my pet-proofing program, but by following the basics and adjusting them to your particular situation you can have that new family member obeying you like a fresh-faced juvenile delinquent on his first day at boot camp. That's when you bring the rest of the family into the program and promote the pet to the role of instructor.

And finally, both canines and felines should be taken out for fresh air and exercise at least twice a day. Remember, dogs go on a leash, and generally speaking, cats prefer the Lincoln Town Car.

# The Dog Rules
### AS THEY APPLY TO THE FURNITURE

It's important to remember that in the beginning furniture consisted of movable receptacles like chairs, tables and beds, created for the comfort and convenience of early human beings. Never was it intended for pets. So …

### RULE 1

Dogs are not allowed on or near furniture of any kind, at any time.

### RULE 2

Okay, the dog can jump up on your lap while you're on the furniture but that's it.

### RULE 3

Okay, the dog can jump up on the furniture if he's got something important to show you, like the stuffed Easter bunny he stole from the kid next door.

### RULE 4

Okay, the dog can get up on the *old* furniture but not the *new* furniture.

### Rule 5

Okay, the dog can get up on the new furniture until it *looks* like the old furniture, and then we'll sell all the bloody furniture in the lawn sale with the doghouse and the cage in a three-for-fifty-bucks trifecta.

### Rule 6

Since it's spelled with a "c" and not a "k," the dog is not allowed up on the BarcaLounger.

### Rule 7

Okay, the dog can sit in the BarcaLounger but only if it's in the upright position.

### Rule 8

Okay, the dog can gear down and put the BarcaLounger in the "sleep" position but only if he's really tired.

### Rule 9

Okay, the dog can shift into the "rocker recliner" position but never the "swivel glider" position.

### Rule 10

Okay, but if the dog activates the "swivel glider" mechanism and he gets sick, he cleans it up himself.

# Did She Say "Food Aggressive"

# or "Food Obsessive"?

THE MANDATE OF THE NORTH AMERICAN Border Collie Rescue Network Inc. is to find good homes for this black-and-white breed of herders, dogs that are stray, abandoned or unwanted.

Somehow I ended up with a bastard. Really. And I mean that in the nicest possible sense. There's more than a little Australian shepherd in Jake, which makes him bigger and so very much calmer than your average Border collie who will play fetch the tennis ball until the sheep have to come home ... on their own. Jake will lie on the floor of my office for hours, allowing me to get my work done before it's finally time for his walk. After two minutes of sitting, a pure and frenzied Border collie would figure out how to use the fax machine and send a complaint of neglect to the Border Collie Rescue Network in New York.

It would be easy for me to say that Jake was once all those things—stray, abandoned and unwanted—but it wouldn't be entirely accurate.

Stray? No, he had never actually been lost and on his own. He was lost and always in somebody's care.

Abandoned? No, not exactly. The couple who disowned him did contact the rescue organization rather than dump him in the dog pound.

Unwanted? Well, here's where it gets a little tricky.

You see, Jake would most certainly have been a "wanted" dog except in Canada we no longer hang posters of criminals on the walls of our post offices.

And he most definitely would have had a place to stay, except dogs, much like our young offenders, cannot be locked up in jail.

And when they refer to Jake as a working dog, they must mean the time he spends doing community service while out on parole, because you see ... THIS DOG IS A THIEF!

I'm sure if I did some genealogical research on the shepherd side of his family it would lead directly to that band of British criminals who founded Australia. I'm guessing this dog is directly descended from a bold little bitch named Ma (The) Barker. (Again, I mean that in the nicest possible sense.)

Jake steals food ... with great regularity and skill.

Under category #9 of Jake's evaluation report, "Dog's temperament," his last foster-home owner and guardian wrote, "Very friendly but food aggressive."

No problem, I thought. He's just a guy who likes to get up in the middle of the night for a snack. Great! I hate eating alone.

I had no idea that "aggressive" included disguising himself as a food bank employee and organizing a charity dry-goods pickup at the local supermarket.

"Aggressive?" I believe she meant "obsessive."

Jake can hardly sleep on Sunday nights knowing Monday is

garbage day on Sunset Bay. So far, I've been blaming all the ripped-up garbage bags on skunks and raccoons, but the neighbors are getting suspicious. Skunks and raccoons don't eat wrapping paper.

Many times I've come home to a house that looks like a room at the Ramada Inn the morning after an alternative-rock band partied there the night before.

Jake opens cupboard doors, knocks the top off garbage cans and hauls anything remotely edible onto his blanket, where he consumes everything, along with some of the packaging.

One marinated T-bone steak, last seen wrapped on a plate and sitting in the sink, turns up as a bare bone and pieces of cellophane on the blanket when I get home at five from the library.

And Jake's so happy to see me, he smiles amid the scraps and garbage on the floor around him. When the evidence is gathered and brought to his attention, his expression alternates between fear

and great relief, and the smile returns to say: "We've been burgled, Bill, but thank God all they got was food!"

Alarmed, I tracked down Jake's original owner and found out that although man's best friend may be Jake, Jake's best friend is the stomach pump. Unlike his previous owner, I've not yet had my medicine cabinet broken into or everything in my toilet case removed and devoured.

Nothing on the kitchen counter is safe; baked goods once kept on the toaster oven are as good as gone the moment I leave the house. I'm talking entire loaves of bread and muffins still in their cooking wraps. It's a shame the name "Jake" is registered, I thought, because "Hoover" fits him so much better.

To keep my house from looking like an accident at the recycling plant, I have resorted to placing mousetraps under sheets of newspapers in strategic locations throughout the kitchen. The idea is that when touched by a trespassing paw, the noise of the trap snapping will scare the larceny out of him, but the paper will keep him from getting pinched. Although this sounds a little cruel, I have to say it's working.

On two occasions, in furtive, middle-of-the-night food raids, the traps have slammed shut and scared the living daylights out of Jake. Actually, my screams of pain scared the living daylights out of Jake, but the sheets of newspaper did save me from requiring stitches.

Sometimes dogs are not all that smart. Sometimes we just have to lead by example.

# The Dog Rules

Certainly there is one room in the entire house free from
the distraction of the family dog.

Hence the den is off limits. No admittance. No exceptions.

You don't sit in his spot at the window and bark;
he doesn't kick back in your TV room.

### Rule 1

Except if it's a dog movie that's on TV like *101 Dalmatians* or
*My Life As a Dog* or *Straw Dogs*. But not a movie that's
described as "a real dog" like the ones Madonna makes.

### Rule 2

Okay, the dog could come into the TV room and watch
educational dog programming like *Nova* or *The Nature of
Things* or even reruns of *The Wild Kingdom*.

### Rule 3

The dog must leave the TV room if he voices his approval or
disapproval by barking, growling, booing loudly or, in the case
of an Adam Sandler movie, lifting his hind leg to the TV screen.

### Rule 4

Okay, if Marlin Perkins is securely locked in his safari Jeep in a
Wild Kingdom episode, and his young partner, Jim, is, as usual,
outside the vehicle and being mauled by two mountain lions,
then the dog can bark once out of fear for Jim's life.

### Rule 5

Okay, twice to warn Jim about the alligator coming out of the swamp behind him as Marlin narrates the unfolding drama from the driver's seat while having a light lunch of hot tea from a thermos and sandwiches with the crusts cut off.

### Rule 6

The dog is not allowed to go behind the TV set to sit in ambush at the spot where the small domestic cats exit from *The All Pets Network*.

### Rule 7

Okay, the dog can go behind the TV set to retrieve his "teddy" but only during the commercials that feature polar bears drinking Coca-Cola.

### Rule 8

Okay, the dog can sit in the TV room, but he's not to get his paws on the remote.

### Rule 9

And okay, if it's an Adam Sandler movie, the dog can join you in lifting a hind leg toward the screen, but only as a dry and symbolic expression of disapproval.

### Rule 10

Okay, the dog can watch whatever he wants but he does not have the right to vote anybody off the island.

# CHAPTER NINE

# *Great Expectations—Dashed*

**A YEAR AGO, IF DIONNE WARWICK** had told me I'd spend the first 15 minutes every day standing in a field watching a dog have a bowel movement, I'd have canceled my subscription to the *Psychic Friends Network*. (But you knew that, didn't you?)

First, I grab my newspaper at the end of the driveway and remove it from the blue plastic bag. And read the paper? Are you kidding? Who has time to read when you own a dog? No, the plastic bag that the paper comes in serves as Jake's pooper-scooper, a highly underrated feature of the *Toronto Star*.

And there I stand at eight each morning, like some kind of perverse cheerleader: "Oh, that's a good boy! C'mon, Jake, you can do it! Jake is gonna feel a lot better! That's it! That's it! Oh, good boy! Jake is such a good boy!"

And people are driving by on their way to work with real purpose to their day, some to construct buildings, others to operate on patients, and I'm standing half in the ditch trying to get a dog to go "number two." And when some of the people driving by wave or peep, I pretend to be reading my morning paper like that's not really my dog squatting in the field taking a public poop. No, I just happened to be reading the newspaper beside a rural road when that stray incontinent dog wandered by.

I tell you, it's singularly shameful. If I had pom-poms, I'd be shouting: "Jake, Jake—he's my man! If he can't do it, nobody can!"

In *PETS Magazine* I read that a dog is less likely to do his business quickly and efficiently if he's on a leash and sees his owner watching him.

I have no problem letting him go to, uh … go, but there's no way I'm going to start disguising myself as an uninterested postal worker just so Jake can feel a little less embarrassed about answering the early-morning wake-up call of nature.

Instead I turn my head and read the headlines out loud, assuring him that I could care less about what he's doing. Until it's time to clean up. Then his business becomes my business, which is why we've named that field after a section of the *Star*—the Business Section.

Mission accomplished. I feel kind of proud, which might indicate that the goals I've set for myself in this life are not real high ones.

A lot of guys start their day with a run or an aerobics workout to keep fit and stay healthy. Not me. I start my day retrieving a little present from my dog. "A little present." That's how dog people talk, in cute little euphemisms. Also, it tells you how easy we are to buy for at Christmas.

Then it's into the kitchen for a little breakfast. His, of course. I make instant coffee for myself so I have more time to: a) measure one cup of lamb and rice pellets into his bowl; b) add a heaping spoonful of leftovers, usually from his Lucy's Café doggie bag; and

c) stir everything with an ounce of olive oil. My muffin can be stale, but this morning mix of dog and people food has to be just right, or he'll take a step back from the bowl and stare at me until it is.

Exactly how did this happen?

Somehow I've become this glassy-eyed, servile, dog-cult member.

For a woman, a dog's charm is like the fatal attraction of the classic handsome bad boy. Say, Paul Newman in *Hud*. They're smitten but remain somewhat cautious. But for a man, the charm of a dog is pure emotional quicksand; take one step toward returning their boundless love, and you're up to your neck in your dog's world for life.

It is nothing less than sheer delight that your dog expresses upon seeing you come home—the tail wags so hard the whole body moves with it, the crinkled nose produces a smile, and the muffled noises of nervous glee escape through dog-breath kisses. If scientists could reproduce this in a bottle, the drug ecstasy would have no popular appeal whatsoever. But it would be no less addictive.

In order to receive the same warm welcome from a woman, a guy must have just returned from a very long journey, like from space or from a foreign war. Yet your dog is this thrilled to see you if you just returned from the store with a quart of milk.

You hear a lot about the love of a dog being "unconditional." And it's true because a dog knows only how to express feelings with his heart. His brain, though highly developed for other endeavors, never enters into the emotional equation. Human love is tampered and twisted around by the brain, which calculates the pros and cons, the advantages and disadvantages of love. Brainless love knows no failure, which makes you wonder why Tommy Lee and Pamela Anderson can't seem to make it work.

Guys relate well to unconditional love because we don't have to think too much. If a woman thinks about anything for any length of time, she'll find a flaw and demand a refund. If a man has to think about anything for any length of time, he'll get a headache and have to lie down.

Having a dog makes people stupid and therefore closer to the elusive and near-perfect state of unthinking love. Which is great because you begin to express feelings for your dog the same way his feelings come to you, unfettered by rational thinking. No calculations or gain or pain.

Hence, being a dog fool makes a man a much better person. But only toward the dog, of course.

The woman in your life still thinks you're an emotional midget whose idea of romance happens annually on February 14 and involves dinner, flowers and the expression: "Holy %@#*! Do you know how much cards cost these days?"

Presiding over a Creature Service for pets of every kind in Vancouver's St. Andrew's Wesley United Church, Reverend Bruce Sanguin hit the old nail on the head:

> "From these pets we have learned many lessons, such as their natural beauty as Your creations, and their unembarrassed requests for affection. In loving and caring for these wonderful friends, we found we were lifted out of our own needs."

And there it was—the solution to man's self-absorption: yanked out of our perceived needs and plunked down in the middle of reality.

When they say, "Get a life!," I think what they're really saying is get a dog. Or a cat. Or both.

It's almost one year since I've had Jake the Border collie/Australian shepherd who was supposed to be a "working dog," but it appears he's taken a very early retirement. He's a "playing dog," and if the game of "stick" ever catches the beady eye of a sports agent, the NFL—National Fetch League—will be created, complete with a network television contract. *Then* Jake will turn pro.

Things are different today than they were when I first brought Jake home from his foster home, where his previous pals do not write to him for fear he'll mistake the return address on the envelope as an invitation to visit them.

In the first few days I had Jake, I must have washed my hands twenty times after playing with him. Not having had a dog since I was a kid, I couldn't believe how slobbery and smelly he could be. And he smelled just like a dog. And he probably had dog germs too. That was then. Now, not only do I not wash after every petting, if Jake doesn't give me at least a dozen wet kisses in the course of a day, by bedtime I'm a little put out. I think I've somehow offended him.

It's amazing—when a dog first comes into your life, it's sit, paw, speak, roll over, and within six months, he's eating at the dining room table and writing checks on your personal savings account.

Exactly how does this happen?

One minute you're trying to teach him how to fetch, and the next minute you're letting him drink from your mug of beer and giving him a couple thousand dollars to go visit his sister in Melbourne, Australia.

On the first day I had Jake I tied him to a tree, thinking if this worked I would build him a nice doghouse where he could cover most of the front yard on a long, sturdy lead attached to a runner with pulleys. Ha! Now I feel fortunate if I can nudge him to the inside of the couch so I can put my feet up to watch TV.

One way I rationalized getting him in the first place was the security factor. I could use a good guard dog around the house. Right. Today Jake would bound to the door with the pink Easter Bunny in his mouth even if it was Hannibal Lecter.

I imagined us walking in the country, the dog heeling, sitting, obeying my every command. Now, after a typical walk and as a matter of courtesy, I get an extension ladder and help the neighbors retrieve their cats from trees and telephone poles.

In the beginning I believed I was getting a companion. I wound up with a roommate that snores and has no visible means of support. I was hoping for a noble beast and I ended up with a hairy brother-in-law.

How in the hell does this happen?

Dogs? Con artists? Thank God they don't have a larcenous bone in their body; otherwise we'd be taking out ads in all the local papers declaring that we are not responsible for any debts incurred by our dogs. But we'd include cute little pictures of them, of course.

# The Dog Rules

As They Apply to Social Interaction

Your dog is the social extension of yourself and as such bears the responsibility of practising acceptable behavior and common manners. That's why ...

### Rule 1

A dog will never jump up to greet a woman ...
unless she's wearing white.

### Rule 2

A dog will never jump up to greet a man ...
unless he's sitting behind the wheel of his brand-new car.

### Rule 3

During a party and with the clothes hamper full of three weeks of laundry, a dog will always grab either white underwear or a black bra to run around the house with.

### Rule 4

A dog will always nose the crotch of the person you're trying hardest to impress.

### Rule 5
A sick dog will always throw up between 3:00 a.m. and 5:00 a.m. just inside the door, beside the mat.

### Rule 6
A dog passing gas will always be under the table where your guests are sitting.

### Rule 7
When you pass gas, that dog will be in another county.

### Rule 8
Provided that his water dish is full, a dog will always drink out of the toilet bowl.

### Rule 9
A dog will never bark at a man in uniform unless it's a postie holding your check or a U.S. Customs officer trying to decide whether or not to pull you over.

### Rule 10
If you had the entire state of Montana in which to walk your dog, he'd pick the park ranger's lawn to poop on and the guy would be home, armed, and gazing out the window at the time.

# CHAPTER TEN

# *Dogs Drinking Beer*

BIRTHDAYS ARE SPECIAL AND REQUIRE SOMETHING a little extraordinary by way of celebration.

Your birthday? No, don't be silly. You'll be lucky to get an age-unfriendly card along with a scratch-and-sniff cologne sample from the center of an old copy of *Vogue* magazine.

But when it's your dog's birthday, well, get the Kodak, Marge, this one's goin' in the scrapbook.

So celebrating Jake's fifth birthday on the day of New Year's Eve 1997, two of us strapped on overnight packs and Jake came along for the walk. Monica Rose, my friend of many years, my photographer when I travel-write and a woman Jake loves more than me, is a walker. A good tennis player, a bad editor, but a long-distance walker all the same. (Sorry, I'm fifty-three years old and whenever I refer to her as my girlfriend, I feel like I just asked her to wear my high school fraternity pin.)

So I called ahead and reserved a room in a motel that accommodates dogs, and the three of us set out on a crisp and sunny winter's day for Dunnville, twenty miles west along Lake Erie's north shore.

We left at 11:00 a.m., allowing six hours for the entire trip including a stop for lunch, which with the drinking water accounted for most of the weight in the backpacks.

Now, at this point, the farthest I'd walked with Jake along this route was about five miles out before backtracking home. Not in the best of shape in those days, he was exhausted when he returned.

You can't believe the look on a dog's face at the ten-mile mark of a one-way walk when he's only been half that distance away from home before and he still remembers the fatigue.

Every so often he'd stop, look back, look ahead, look back, then look directly at me with that incredible inquisitive expression like: "You got a back-up plan, right? Public transit? John the neighbor? Somebody's driving us home, am I right on this one?"

And I did have a plan: to be in Dunnville before six o'clock when the liquor store closes. Hey! It's also New Year's Eve.

Walking is a wonderful way to travel. You don't make great time, but you see everything there is to see, like other dogs, big protective farm dogs not entirely happy about you venturing close to their domain. Most can be fended off with a low, threatening voice, a gross subhuman sound, the kind that turned Linda Blair's head all the way around in *The Exorcist*.

The others, spoiling for a fight, need three rocks, each the size of a fist. The first is an early-warning signal, thrown hard and well in front of the approaching troublemaker, making the maximum amount of noise. The second one, if necessary, is thrown directly in the path of the oncoming animal so that is skips and rolls to him or very close. The third, gripped in the hand of a cocked arm, accompanied by a voice as menacing as any growl, is launched full force to the area in his immediate path. If you can hit a metal sign or a tree, it's probably the last rock you'll need.

Of course, it's not the dog's fault. A dog with a nasty temperament running loose is an animal of low intelligence and shabby self-esteem. However, if you halve the intelligence quotient and drop the self-esteem ranking by a good forty points, you've pretty much got the profile of the owner.

A dogfight—and I have been witness to some very vicious bouts—

is an unnecessary and demeaning act committed by one, and sometimes two, very irresponsible people.

In all savage attacks and fights involving dogs, I think serious attention should be given to healing and rehabilitating the animal while taking good care to humanely put down the owners. (Sorry, but I've really had it with animal owners who believe their obligations end with food, water and shelter.)

Now, normally I'd spend New Year's Eve on some cheap but exotic trip to Cuba or Mexico, but no, now that I have a dog who is deemed to be too good for the common boarding kennel, I find myself having lunch at a frozen picnic table in Lowbanks and fine dining in a place called Dunnville, a town whose mascot is—I'm not making this up—a mudcat. (A mudcat is a catfish that is often chased, underwater of course, by a dog-fish—see The Dog Dictionary.)

And I asked myself as I beat a frozen orange on the table to shatter its peel: "Do dogs change your life?" No, not really. They rule it. Fortunately, they're very benevolent dictators; otherwise we'd all be living in labor camps run by large, armed dogs wearing helmets and sunglasses.

So, to keep warm, we kept moving, walking along the winding country road that hugs the Lake Erie shoreline past old rural dwellings, with farms to the right of the road and the mostly boarded-up cottages, many owned by Americans, on the left side by the lake. About every fourth or fifth lakeshore home is a winterized cottage, larger and looking lived-in, occupied by a brave but not too bright Canadian known as "a year-rounder." Unlike the "seasonal" or "summer people," year-rounders spend their winters wrestling frozen water pipes, sleeping by the fire for days during ice storms and moving to higher ground when the lake rises to kill them in the middle of the night. I'm proud to be a year-rounder, and now, so is Jake. Comfort, we believe, is for city people and sissy dogs. We're in it for all four seasons and this is Wainfleet, dammit, where men are men and women can barely keep from laughing.

Past Lowbanks lies—wait for it—Highbanks, reddish cliffs of clay and sand that fall sharply to the beach, and on this day are snow covered and walkable.

There is something bleakly beautiful about a summer retreat in winter: The trailer park empty and perfectly still except for the banging door on the open, outdoor refrigerator or a loose cable that repeatedly whips a flagless pole. The mast of a sailboard somebody forgot on the beach serves to measure the thickness of the ice, and the animal tracks in the fresh snow around the cottages and sheds render the image of a wildlife colony operating beneath the floorboards and down the chimneys, come to life after the summer people have gone home.

And it went on like that for twenty miles, a meandering winterscape of cottage country and farmlands, asleep for now, in a coma of cold. And Jake enjoyed each and every part of it with the lifting of a back leg and a trademark yellow spritz of approval.

We made Dunnville and the liquor store with fifteen minutes to spare and a good selection of champagne still on the shelf. We confirmed our dinner reservation at Molly Ann's Café and then walked through the business center to the far side of town. The Riverview Motel with its warm and clean units overlooking the Grand River accepts both dogs and, as we later learned, loud guys who walk on all fours at four o'clock in the morning.

When we arrived in the room, I was in desperate need of two

things. Okay, three, but don't get too far out in front of me here. (And if it were just as simple as lifting a leg upside a tree in broad daylight—don't you think guys would ?!?)

I needed, we needed, we all needed a very hot shower and an extremely cold beer. Jake passed on the shower, preferring to flop onto a throw rug next to the radiator, and from this position gulped tap water from a plastic bowl.

It was while I was having myself a well-earned steamy and frothy good time that an unfair thought crept into my mind—tap water?

In five hours it would be 1997 and in five hours and one second Jake would officially be five years old and he's scrunched into the corner of a motel room drinking recycled water from the Grand River? I don't think so.

I grabbed a still-cold can of beer from my backpack and replaced the water in Jake's bowl with what I imagine he'd heard many times referred to as a "brewski."

Because, as I quickly found out, this dog *knew* beer the way a rummy knows cooking sherry.

Jake took one whiff of the sudsy bowl, gave me a doe-eyed look of love and eternal gratefulness and then lapped back that beer like he once served an apprenticeship with three guys named Bob on bar stools at the Belmont Hotel in Port Colborne.

I could tell by watching him that Jake was no amateur imbiber. No sir, because after he finished his beer, after he licked every bubble from his plastic bowl, he belched and then crushed the can on his forehead.

Yet another reason surfaced on why this dog had been bounced around from home to home. He'd probably figured out the twist-top system on the caps of beer bottles.

After a failed attempt to con me out of my glass of beer, Jake made three wide circles in the middle of the motel room, and he didn't wake up until the following year.

As we dressed for dinner, I wondered if beer was bad for a dog.

I'm sure the occasional splash at the end of a special day would be okay, but I could also imagine scenarios where it could get out of hand, like a normally docile cocker spaniel showing up at the vet's with beer on his breath, picking fights with all the cats ... or a pack of greyhounds running down a Saint Bernard and stealing his cask of brandy to mix a bunch of boilermakers.

A table of German shepherds wearing lederhosen and drinking draft beer from steins until even the ugly little pug in the corner looks good to them at closing time.

Yes, I could see when it might be a problem, but for now, as Jake enjoyed a deep sleep and the occasional snore in a happy heap on the carpet, "Beer been berry berry good to him."

After a great dinner and a few pub stops along the way, Monica and I returned to the Riverview to find Jake safe and sound and silent. No barking is the rule when dogs are doing sleepovers.

Apparently, while we were gone, Jake had drunk the rest of my six-pack and polished off part of a bottle of Armagnac. When we walked in, he was cuddled up at the foot of the bed asleep while pay-per-view was showing *101 Dalmatians Meet 35 Huskies in Heat*.

Anyway, it's one of the few times I've given Jake a beer, and it's more a matter of economics than ethics. He's proven to be quite a thirsty fellow, and his habit could get expensive.

You may not think much of this idea, but drinking beer is certainly one experience I could never share with my cats.

# The Dog Rules

The Mardi Gras for Mutts parade in New Orleans. Okay.
The Yuppy Puppy Dog Daycare Centres in Vancouver. The
Pooches Without Partners evening at the Three Dog
Bakery, also in New Orleans. Camp Gone to the Dogs in
Putney, Vermont, featuring square dancing. Theme package
vacations like Houseboating with Hounds and RVing with
Rover. Okay. But the $13,000 U.S. Dordogne Doggie Walks
in Southern France, which includes return tickets for you
and your dog on Air France's Concorde? Really.

### RULE 1
The dog must remain in his carrying cage in
the pet cargo hold at all times. It's
a U.S. Federal Aviation Administration law.

### RULE 2
Okay, to be perfectly fair, the pets can join the passengers
for the safety briefing. But, they must return to their cages
when the seat-belt sign goes off.

### Rule 3

No barking. I told you, when the oxygen mask drops from
the ceiling I will put it over your mouth. Okay?

### Rule 4

I know, I know. Paying a flight attendant to show you
how to operate a seat belt is pretty stupid.

### Rule 5

Quit looking at me like that! There was no window seat!

### Rule 6

Don't even think about it. We're at 38,000 feet
and that cat is scared enough.

### Rule 7

I don't know. I guess too many people were allergic
to peanuts. Just make do with the pretzels.

### Rule 8

Okay, the dog can stay for the meal but not the movie.

### Rule 9

Don't get upset, maybe she just doesn't
like dogs. Do that thing where you smile
and stick the tip of your tongue out.

### Rule 10

No, I don't think there are any female dogs on this flight.
What? The Mile-High Kennel Club?

# CHAPTER ELEVEN

# *Man's Best Friend or*

# *His Worst Embarrassment?*

"A DOG IS MAN'S BEST FRIEND!" We've been hearing this fabled phrase forever.

However, the popular little proverb may not be exactly true and it's not exactly what the man said. It's actually a composite quote from a speech given before the U.S. Senate in 1884 by George G. Vest who said: "The one absolutely unselfish friend that a man can have in this selfish world, the one that never deserts him, the one that never proves ungrateful or treacherous, is his dog ... He will kiss the hand that has no food to offer ... When all other friends desert, he remains."

This glowing account of canine character would least describe my dog, Jake. First, Jake is no hand-licker. He's a relentless sneak-attack kisser who nails you when you least expect it, like first thing in the morning. Yawn and you're as good as Frenched.

Second, Jake goes where the groceries are. "No food—no fare" is this dog's motto. And he could find a table scrap in the middle of the Sahara.

But never mind Jake, as hard as this dog might be to ignore.

Let's focus for a moment on a bright white cottage, a dog, one man, one woman and Wainfleet. Hey, wait a minute! According to the 1998 community census—that *is* Wainfleet!

John and Nancie Grant are friends and neighbors who live seven cottages east of me. They own a large, handsome, seven-year-old, pure-bred husky by the name of Nukey. He is a model of his wolfish breed, looking for all the world like he just stepped out of a Robert Bateman painting.

Let's just see if Nukey fits the accepted popular image of man's best friend as laid down by George G. Vest.

Not long ago John and Nancie came to my place for dinner on a Saturday evening.

At precisely eight o'clock, the time the Grants were supposed to arrive, a car pulled into my driveway with the lights out. Surprise visitors, I thought. Bad timing and strange as well.

I was puzzled to see John and Nancie emerge from the car and quickly scurry into the house.

"What's with the car?" I asked as I took their coats at the kitchen door.

John and Nancie looked at each other nervously, embarrassed, like the parents of a kid who'd just landed the lead in a *Crime Stoppers* episode.

"Ah well," John began in a halting confessional tone. "Nukey's on his lead out front, and we didn't want him to see us walking to your place because he'd think we were visiting Jake. That would make him furious. He'd sulk for days." Obviously, I thought, this has happened before.

So, to deceive this dog, who is so genetically close to a wild animal wolves keep dropping by the house trying to get him to defect, these two grown-ups, both of whom have university degrees, drove past my house as if they were going shopping in Dunnville, then made a U-turn once they were out of Nukey's sight, came back with the headlights dimmed and shut off the engine as they coasted up my driveway under the cover of darkness. Burglars use this same technique but don't look half as embarrassed. They did everything but put black shoe polish under their eyes and creep slowly to my kitchen door wrapped in cedar boughs.

"It sounds crazy, but it'll avoid an ugly scene at home," added Nancie.

Let me assure you, these are two mature, conservative, normal human beings. Although I've always wondered about the trapeze and leather masks in their guest house.

After additional questioning it turns out John and Nancie sleep most nights in severe fetal positions to make room for Nukey at the foot of the bed. And when it becomes unbearable, cramped or too hot, instead of ordering him to get down, they convince him Jake is outside in the yard stealing his buried bones again, and the dog is so jealous he goes and sits for hours by the window to keep watch. You're probably thinking, man, this one's not too bright, but, in fact, John is a retired public school principal.

Every Halloween John and Nukey are the hit of the neighborhood's trick-or-treat parade.

Nukey gets noticeably excited when he sees John wearing his latex werewolf mask complete with hairy hands and feet. Nukey thinks John is actually his long lost alpha uncle risen from the pre-domestication gene pool.

Somehow it's difficult to take a husky seriously when he's wearing

84

Christmas reindeer antlers on his head and a flashing, back-up reflector light Velcroed to his bum. He looks like a rescue dog that spent too much time at the Chernobyl meltdown site.

Merrily they march along, Nukey gathering doggie treats in a plastic bag and John accepting small sweet offerings that are rum-based with ice.

Funny, as they go farther afield, fewer and fewer doors are opened to them, and more and more porch lights go out just as they start up the driveway. John convinces himself it's just the lateness of the evening. Nukey figures it's John.

Not nearly as embarrassing but no less arresting is the sight of John sailing along the shoreline of Lake Erie at the rudder of his sixteen-foot catamaran and Nukey at the helm wearing a bright yellow flotation vest and a blue captain's hat. It's the kind of scene Norman Rockwell would have painted, if Norman Rockwell had had a serious drug habit.

Jake sometimes swims out because he can't believe what he sees from shore and Nukey sails right on by him with that "I'm king of the world!" arrogant look on his face.

There are days when both Jake and I think they could both use a really hard iceberg.

And when John and Nancie go away, Nukey doesn't go to just any kennel, oh no. They searched and researched until they found a private kennel, in a rustic setting, that borders a farm where Nukey could enjoy his hobby: Sitting and looking longingly at cows. Probably sitting and imagining how their prime ribs would look medium rare with a nice mushroom gravy.

Bob Scott calls his kennel Pet Care Camp. John and Nancie can get a huge howl out of Nukey by chanting, "Cow Camp—Nukey! Cow Camp!"

When Nukey is not sitting in a field watching cows, he's home in front of the television set watching reruns of those PBS *Nova* shows that contain explicit audio representations of wolves howling, which John and Nancie tape for him and play back on the VCR when the dog is a little depressed. (Funny, but the last time I was down in the dumps they told me to quit feeling sorry for myself.)

Then John—and yes, under cross-examination he has admitted getting inside Nukey's doghouse and having a conversation—gets down on the floor in front of the television set and starts howling at the moon with the dog. And of course, when the neighbors complain about the noise, John pins it all on the dog, scolding him and sending him to bed without his favorite television show.

Oh sorry, did I say "dog?" No, Nukey is always referred to in the Grant family as "the boy." So if you knew John and Nancie, but never went to their house, you'd think they were pretty cruel people chaining their son outside, feeding him ham bones and quenching his thirst from a garden hose.

My only question is: When they have to call the vet and Dr. Thorne returns that call, exactly which sick puppy at the Grant residence picks up the phone?

What worries me most is that John and Nancie are allowed to operate motorized vehicles and use prescription drugs.

Oh, and did I mention John and Nancie have two fine daughters whose names are, according to Nancie: "Oh, you know, the short one with the dark hair who's with Air Canada, and the blonde who's the computer whiz?"

Man's best friend? I don't think so. Life's most embarrassing moments with mutts? Probably.

I mean, when you add up all the bones Jake has stolen from Nukey over the last two years—dogs aren't even *their* own best friends!

# The Dog Rules

For reasons of health, hygiene and a quiet, undisturbed night's sleep, the dog should remain just outside the door-way during the night. Because ...

### RULE 1

The dog never sleeps on the bed. Period.

### RULE 2

The dog can sleep on the floor on either side of the bed or at the foot of the bed provided he's on his monogrammed dog bed that cost $150 plus shipping from L.L. Bean, which he hates.

### RULE 3

Okay, the dog can jump up on the bed just before you go to sleep or in the morning when you wake up, but just for a short visit.

### RULE 4

Okay, the dog can sleep on the bed but only at the foot of the bed provided he's on his blanket.

## RULE 5

No, the new sheets and covers are not to be selected
according to how they color-coordinate with the dog.

## RULE 6

The dog never sleeps alongside you like
another human being.

## RULE 7

Okay, the dog can sleep alongside you, but he's
not allowed under the covers.

## RULE 8

Okay, the dog can sleep alongside you, under the
covers, but not with his head on the pillow.

## RULE 9

Okay, the dog can sleep alongside you, under the
covers with his head on the pillow, but if he snores,
he's got to leave the room.

## RULE 10

Okay, the dog can sleep, snore, fart and have all the
nightmares he wants, but he's not to sleep on the
couch in the TV room, where you now have to sleep.
That's just not fair.

# CHAPTER TWELVE

# *Dog Evolution—*

# *From Darwin to Doggie Bags*

**I WAS SITTING ON THE BREAKWALL** the other night, the waves setting a pensive rhythm to the quiet lakeside scene. The orange glow of the fire revealed a middle-aged man dropping potato chips at his feet, where his dog alternated between eating the salty treats and taking the occassional sip from the man's pint of beer, and I began to wonder how we, humans and canines, ever got to this point. The point at which all owner/dog interaction belongs in the comic strip of life.

What the hell happened to us?

I'm told that thirty million years ago the first doglike creature, Cynodictus, appeared on Earth. He hated his name—Cynodictus—and disappeared under the bed for 29,700,000 years. (To put that in perspective, thirty million years is the total amount of sleep the average cat gets in one lifeline.)

Cynodictus developed into what we now know as a wolf almost 300,000 years ago. Farley Mowat was on the scene and covered this event for the old *Toronto Telegram*, eventually publishing a book on the subject entitled *Never Cry Wolf*.

About 40,000 years ago, wolves and humans found themselves in a daily competition for food, hunting and consuming small animals and then sharing the same green grass whenever they ate a bad rodent.

Wolves quickly realized that wherever early man settled, bones and skins would be strewn nearby. Rather than hunt for themselves, the wolves began to dine on man's discards. The wolves called these tasty remnants "leftovers." It was the beginning of the Monday-night casserole.

It was widely believed that 12,000 years ago, wolf cubs were tamed and bred over many generations, and that eventually these wolves became domesticated dogs. Today it's hard to believe, but apparently dogs were originally domesticated for specific reasons other than how cute and stupid they are and how much they will love you and kiss you even if your own mother thinks you're a creep. Today, nobody can remember what those reasons are.

But the question remains: How in the name of evolution did a fierce wolflike carnivore come down from the mountains thousands of years ago and wind up sitting under my lawn chair in Wainfleet, Ontario, eating BBQ chips and drinking cold Bass Ale until finally he belches out loud, and I have to say: "Excuse me?"

When the beast that once ran in packs and had slow-footed cavemen for lunch winds up sleeping next to you in bed and growls because the reading light is keeping him awake—sorry, but this is not exactly a quick hop over the gene pool pond.

I don't care if Charles Darwin is your travel agent, this is still a helluva trip.

Recently I sat riveted watching a CBC television documentary by David Suzuki entitled *Man and Dog—An Evolving Partnership*, with expert guest professor Ray Coppinger, a dog behaviorist. (Unfortunately Jake missed this show. He was in the other room at the time ... doing my taxes.)

Can you imagine being a lifelong *dog behaviorist*? At the end of an enduring and successful career when you sit down to write your memoirs, there would have to be one entire volume on *Drinking Out of the Toilet*.

Professor Coppinger believes that every dog on Earth today can be traced back to a wolflike creature that lived approximately 12,000 years ago. Not only have archeologists discovered cave drawings that support this time frame, but archivists have even found a depiction of a wolflike tattoo on the left buttock of Dr. Ruth. (Sorry, but even sex should have some sort of age limit.)

"Scavenging" is the keyword in Professor Coppinger's theory of how 100 million wolflike descendants, commonly known as pets today, have evolved to the point where they now watch rented movies and ride shotgun in pickup trucks. They knew!

Dogs knew 12,000 years ago what it took P. T. Barnum most of his life and a herniated disc from shoveling elephant poop to learn. Precisely: When it comes to man, there's a sucker born every minute.

According to the distinguished Ray Coppinger, a Harvard professor who's an expert on "biofilia," or the need for species to associate with other species, we did not seek out dogs for domestication, as was until recently commonly believed. No, those cunning little drooling devils saw us coming, a long way away. And they came looking for us.

Simply put, wild dogs figured out a long time ago that instead of stalking and running and killing their prey in order to survive, it would be a lot easier to follow early man around and live on what missed his mouth. This is true. Then, circling the edge of the man's encampment proved as fruitful, foodwise, as today, when a smart dog patrols the end of the table where the head of the household sits. You'll never see a skinny dog sitting at the feet of a fat guy having supper.

Dogs, then and now, have proved to be superb scavengers, born experts at extracting food from man by cleaning up after his activities. From "cave crumbs" to "table scraps," dog has evolved. Dogs were the first species to switch from a wild hunting carnivore to a parasitic companion by getting close and living close to man in a symbiotic relationship. (Apparently, other wild species of the time had much higher standards.)

So the village dump sites became the dogs' cafeteria, and the discarded entrails of animals and fish became the daily specials for the feral dog that lived on the peripheral of man's home and community, all the while keeping a safe distance from human hand.

It was then and only then that man discovered dogs could be useful in guarding, hunting and herding, thus elevating them to a preferred position in the food chain and domesticating them in the process.

The popular theory that early man adopted wolf puppies and tamed them for his own use is now dead.

First of all, early man, like modern man, wasn't all that smart.

Trust me, early man sitting around the fire after eating blackened ox, quaffing mead, scratching himself and belching without being scolded (this, to modern man, is the period often referred to as "heaven") did not suddenly say: "What I really want is a dog ... to walk, to bathe, to feed, to watch squat in a field first thing every morning."

No, that never happened.

What happened was early man came home drunk one night, and having forgotten to kill supper, blamed it on the dog that lived on the edge of his property. Overhearing this, the dog ran off and in minutes returned to drop a still-warm wild turkey at the mouth of the cave. Early woman, realizing the severe limits of intelligence common to all untamed beasts, threw the bum out and adopted the dog. And there you have it—domestication of dogs as well as the beginning of trial separations.

She taught him to guard the entrance to the cave, hunt for fowl and herd sheep. But it just wasn't the same. He wanted desperately to sleep inside the cave, with her, and the dog. So the big historical compromise was struck.

She agreed to let him move in as long as he quit carousing with the boys and dragging other women back to the cave. And if it didn't work out, he'd leave and she would get half of everything he owned. The arrangement was called "marriage."

His fellow cavemen were flabbergasted. "Can anyone get in on this deal or what?" they asked incredulously and then got in line for the certificates.

The dog stood by silently, watching all this, and slowly but surely taking over the house. Eventually the dog decided to let them both stay if they promised to behave themselves and stop bickering.

Today, even in the world's most expensive restaurants, the chef will send his finest cuisine home to the dog in tinfoil wrapped in the shape of an elegant swan. Nobody, least of all the dog, knows why they do this.

Now whenever I walk Jake, he immediately grabs the leash out of my hand and takes off down the driveway with it in his mouth.

It's just a subtle reminder of how things actually are. Jake feels more secure when the leash is attached, but I hate it because it leaves an itchy, red ring around my neck.

# The Dog Rules

Any communication with your dog while you're away could unnecessarily excite the animal, making him believe you're coming home earlier than planned. Therefore ...

### RULE 1
Do not leave prerecorded, automatically timed messages around the house like: "Hey! Get off the couch, you big lug! I mean it! Right now!"

### RULE 2
Do not make contact with your dog while you're away.

### RULE 3
Okay, a funny postcard addressed to your dog that you'll read to him when you get home wouldn't hurt.

### RULE 4
Okay, a phone call to the house sitter asking that person to relay a nonemotional message to the dog would be okay.

### RULE 5
Asking Bob, the guy who runs the kennel where your dog is staying, to go to his cage and say: "I wuv you, I wuv you, I wuv you!" is embarrassing for everybody involved.

### RULE 6

Okay, asking the house sitter to hold the phone
to the dog's ear while you yell: "It's me, Sparky!
Can you hear me? It's me!"
That can't do much harm.

### RULE 7

Asking the dog to "Speak! Speak! It's me, boy! Speak!"
while the house sitter holds the phone for him is okay,
but you gotta pay her extra.

### RULE 8

No singing on the phone ... by either of you ...
even if it is bedtime in both time zones.

### RULE 9

Phoning home from your airplane seat to
tell your dog you're on your way is ...
simply a case of common courtesy.

### RULE 10

Asking the person in the seat beside you if he wants
to talk to your dog too is just silly.

# CHAPTER THIRTEEN

# The Dog Judge

**IT'S PRETTY SIMPLE:** In today's world, when two people have a serious disagreement over a particular issue—they just sue each other.

However, when a dog and his master ... (I threw that word in to see if you're still paying attention—it's called comic relief.)

Let me try that again. When a dog and his mark find themselves in a situation of dispute, there is no system in place in which a lawyer could mediate and settle the conflict, thereby stripping both of them of every meager possession they have, along with their dignity. (And they call it civil law.)

Sorry, I was married once.

When a dog and owner find themselves in contentious circumstances, there really is no mechanism of arbitration available to negotiate an amicable end to it. Yet there should be. Quick and common-sense solutions to crises are essential to a good and solid relationship.

In a world in which we now have paid mentors, personal fitness trainers and life coaches, I think it's high time we had dog judges.

Invested with the power and authority to settle all disputes between dogs and their owners, dog judges would be like any other court-appointed mediator, except that they would always have a plastic bag under their gowns in case one or the other party in the dispute became very excited and pooped. And they would also make house calls.

A case in point would be an ongoing quarrel Jake and I have over my low, living room couch.

I, commonly known as the plaintiff in this case, have asked, begged, demanded and threatened Jake not to sit on the couch, which happens to be made of chocolate-brown corduroy, an exquisite background for dried mud and dog hair.

Jake, commonly known as the pontiff in this case, because he believes he has a God-given right to this couch, regularly ignores the furniture rules.

This dog will actually get up in the middle of the night, having a perfectly good and comfortable place for a dog to sleep, namely, my bed, and he'll go into the living room and curl up on the couch because HE KNOWS HE'S NOT SUPPOSED TO AND BESIDES, IT MAKES ME SO CRAZY I START YELLING AT THE TOP OF MY LUNGS, IN CAPITAL LETTERS!!!

JUDGE: "All right, that's enough. One more outburst from the plaintiff and I'll clear the living room. I'll ask you to just answer the questions and calm down. Respond in lowercase only. Is that clear?"

ME: "Yes, Your Honor."

JAKE: "Woof!"

JUDGE: "Quiet. You'll have your chance to speak later."

ME: "That's not the worst, Your Honor. I'm holding my own in that battle by putting mousetraps on the couch at night. It's hell on houseguests who sleep over, but those cases are being heard in the *People's Court*. No, the worst is that I'll be sitting on the couch reading the paper, and Jake will come in and sit with his bum on the couch."

JUDGE: "Which is it? Is he on the couch or is his bum on the couch?"

ME: "Well, I'd say he's on the couch."

JAKE: "Woof! Woof!"

JUDGE: "Not now, I said. And get that Bible out of your mouth!"
ME: "And he'd say, technically, that because all fours are on the floor, that he's legally sitting on the living room rug, where he is allowed, and his bum is just resting on the couch."
JUDGE: "Is his bum allowed on the couch?"
ME: "Well, no. Not really ..."
JUDGE: "Have any other bums ever spent time on the couch? Like some of your rowdy tennis buddies after a night of drinking?"
ME: "Judge. I don't think I should answer that because—"
JUDGE: "It's okay. I was there once myself. Proceed."
ME: "Could we get back to the bum at hand?"

JUDGE: "How big is this dog's bum?"

JAKE: "Woof! Woof!"

JUDGE: "Sorry. It's relevant."

ME: "Well, that's one of my nicknames for him, 'Dog with the big bum.'"

JUDGE: "But he's a big dog."

ME: "Okay, then how about 'Big dog with the big bum.'"

JAKE: "Woof!"

ME: "Or what about 'The big mutt with the big butt!'"

JAKE: "Woof! Woof! Woof!"

JUDGE: "Sustained. I'll pretend I didn't hear that last crack."

ME: "So my point is, when I tell this dog to stay off the couch, I think any reasonable person would agree the order covers all dog parts including his bum, irregardless of size."

JUDGE: "Irregardless?"

ME: "Yeah, I want the dog to understand what I'm saying too."

JUDGE: "It seems to me you have to be more specific. Let me see, here it is, in the matter of divorce in the case of *Thomas versus Thomas* ... Are you familiar with the phrase 'Why don't you get your fat ass off the couch and go mow the lawn?'"

ME: "And another thing, Your Honor, if he comes in the house wet and I tell him to get on his blanket, most of him ends up on the blanket, but his front paws always overlap onto the carpet."

JUDGE: "So his paws remain outside the designated area of the blanket?"

ME: "Yes, followed by his head and neck, which flop down on his front paws."

JUDGE: "And then he does what?"

ME: "He sits there and stares up at me in dog defiance."

JUDGE: "And you see this as disobeying a direct order?"

ME: "Oh come on Judge, he's triple dog daring me to move him."

JUDGE: "Wow, a triple dog dare takes double jeopardy to another level entirely."

ME: "Then he wiggles his bum farther onto the floor like: 'Get a lawyer, I've read the Dog Charter of Rights!'"

JUDGE: "Well so have I, and so in this instance I'll have to side with the pontiff. Just as all male human behavior is covered to a large extent under the legal realm of *duplex mensura* or the *old double standard*, the same legal rights are enjoyed by a dog in a canine/human relationship. In dog Latin (see The Dog Dictionary) it's called *juxta satis*, or in plain English *close enough*. Now, unless there's anything else, I'll—"

ME: "There certainly is something else. He begs for food at the table when I've told him not to."

JUDGE: "Does he bark or paw the people at the table or nudge their napkins or anything?"

ME: "No, he just sits there with those big pleading eyes and head cocked a certain way and tongue hanging out."

JUDGE: "Oh hell, I have to do that in order to have sex with the wife. That's not illegal. No, there's nothing wrong with that. Now, I'm throwing out the blanket charge and the begging thing, and I'm ruling that bums are allowed on the couch. Both of you."

JAKE: "Woof! Woof! Woof!"

JUDGE: "What did he say?"

ME: "Close enough."

# The Dog Rules
## As They Apply to the Dining Room

Here's the deal—you don't go near the mat with the water
bowl and the plastic food dish, the dog stays
away from the dining room table. Period.
It's called the separation of the species.

### Rule 1
Okay, the dog can sit at your feet, under the dining room
table, in case you need to tactfully dispose of something
you don't like. Like liver. Or Brussels sprouts.

### Rule 2
The dog is never allowed to actually
sit at the dining room table.

### Rule 3
Unless it's for a gag photo. Then the dog can
sit at the table wearing a party hat and a bib, but
he's not allowed to actually dine.

### Rule 4
Okay, if he's good, and smiles for the camera, he can
have a treat at the table, but not a full-course meal.

### Rule 5

Okay, the dog can sit at the table wearing a party
hat and bib and eat a full-course meal only if a guest
failed to show up for dinner. (I mean, he's
going to inherit the food anyway, right?)

### Rule 6

Okay, the dog can put his paws on the
table, but not his elbows.

### Rule 7

The dog's eating is confined to his own
plate, on his own place mat.

### Rule 8

Okay, the dog can request food from others at the
table by barking softly, but he's not to reach over
and take stuff without asking.

### Rule 9

The person on the dog's right can apply a napkin
to the dog's mouth and chin if he needs it, but
that person is not to pour the dog wine.

### Rule 10

Okay, the dog can taste the dinner wine,
but not first. And not out of the bottle.

# CHAPTER FOURTEEN

# *Say Goodnight, Spike*

NOW, AS YOU AND I KNOW, dogs don't talk. At least not in so many words, like us.

Which is a very good thing because if you've ever tried to call into or out of a house where teenagers live, you can imagine what the telephone gridlock would be like if all the dogs in the neighborhood could ring each other up just to gossip, mostly saying catty things about felines.

No, dogs don't talk.

Six summers ago, long before I got Jake, I was driving my little Miata up to the Gatineaus of Quebec, one of Canada's truly unspoiled splendors, and I found myself being assaulted by one radio talk show after another. The French stations were easier to take because I could understand only half of what they were saying.

(Nobody talks on talk radio anymore—they spit bile at each other through the anonymity of the telephone and the safety of the radio signal.)

So I reached up for my cassette case, which, of course, I'd locked safely in the trunk. I suppose I could have pulled over and retrieved it, but that would have snapped my thirty-three-year-consecutive non-stopping streak on long trips, which is why I can never ever drive to Nova Scotia without a spare kidney in the beer cooler. I like to think of myself as Mr. Non-Stop, baseball's Cal Ripken of car trips.

So I ran my hands under both seats until, sure enough, I found a tape. It was a homemade tape I hadn't seen for years. It's just one song, repeated over and over by Louis Armstrong, entitled "What a Wonderful World."

Mindlessly I drove and half listened to this song until well into the second hour, when one of the lines jumped out at me like a priest in a police lineup.

"What a Wonderful World" is a sweet, short ballad that goes like this:

> *I see trees of green,*
> *Red roses, too,*
> *I see them bloom*
> *For me and you*
> *And I think to myself,*
> *What a wonderful world.*

And I thought to myself, gee, if only I had a rhyming dictionary, I could move to L.A., change my name to Mitch, make my pitch, find my niche, create musical kitsch, and without a hitch be as rich as Marvin Hamlisch. Life's a bitch, but still and all—it's a wonderful world. Especially when Louis Armstrong sings:

> *I see skies of blue,*
> *And clouds of white,*
> *The bright blessed day,*
> *Dogs say goodnight,*
> *And I think to myself,*
> *What a wonderful world.*

Excuse me?

Dogs say goodnight!?!

I don't think so. Dogs don't say goodnight. There's no talking in the dog world. Dogs don't talk. Dogs may whine, whelp, woof, wiggle, pee, prance, sneeze, snort, snuggle, run, roll, jump, drool, pass gas, catch a ball, shed, scratch and sniff—but dogs do not possess verbal skills beyond that of a French mine at a silent charity auction for the hard of hearing.

Dogs do not talk, which is the main reason dogs were never used as double agents during WWII.

This is why, even though he was interviewed twice for the job, Eddie, the dog on *Frasier*, is not Rosie O'Donnell's cohost on her talk show. That, and he took a leak on the set, something only a disgruntled guest had done before.

This is why you will see a lot of dogs with their heads sticking out of moving vehicles, but you'll never hear one say: "For crying out loud, slow down, Slick. I'm tearing up over here."

So, I'm sorry, but with all due respect to Louis Armstrong, dogs don't say goodnight.

*... clouds of white,*
*The bright blessed of day,*
*Dogs say goodnight.*

Don't get me wrong. Dogs communicate, and effectively too. For instance, a dog will sit nervously and stare at you while you're eating, in essence, begging for food. A dog will not, because he cannot, put his front paws on the table, look you in the eye and say: "Yeah, like you need another pork chop and a fourth scoop of mashed potatoes! Have you looked at yourself naked in the mirror lately?"

No, canines cannot communicate, which is why no dog in the history of dog shows has ever thrown a hissy fit in the finals and called his competitor a "%@#*ing prima donna."

And really, if dogs could talk, the last thing they'd say would be "good-night." The first thing they'd probably say is: "You go through with this, Doc, and when I wake up, you better have gotten rid of the idea of ever having kids yourself!"

And the second thing they'd probably say is: "Would you quit putting those blue pucks in the back of the toilet? It's like drinking Zima."

Think about it, if dogs talked they wouldn't have to chase cats. They would just yell things like: "Your mother sleeps with strays in the alley!"

No, dogs don't talk, and in some cases it's a damn shame. I mean, if it was possible and if dogs could talk, O. J. Simpson would right now be coaching a federal penitentiary football team and dating a guy named Snake.

Think of everything your dog has seen you do. See what I mean—

if he could talk you'd have to make him put his paw on a dog Bible and swear him to secrecy.

For guys, one of the dog's most attractive features is that he does not talk. That's why a dog is man's best friend and a parrot is not.

No, I'm sorry, Louis. Skies may be blue, clouds certainly are white, days can be bright, dogs can't say goodnight.

You can imagine how I felt weeks later after ranting about this to a friend, going over the whole case I just made for dogs not talking, and she said something I'll never forget. She said: "It's *dark sacred night*, stupid."

> *I see skies of blue,*
> *And clouds of white,*
> *The bright blessed day,*
> *The dark sacred night.*

Yeah, that sounds about right. Okay, so I made a little mistake. Hey! It's still a *Woof! Woof! Wonderful World* anyway.

# CHAPTER FIFTEEN

# *See Mexico by Dog*

I'M QUICKLY DISCOVERING that not only does a dog take over your house and your car and your life, he also acts as your travel guide while you're on holidays thousands of miles away from home.

Every trip begins when I haul out the luggage, and those eyes, big, brown and bright, become instantly glaring and sad. He flops down on the floor, his head on his front paws, and though his head does not move, his eyes follow every move I make as I pack my big black bag and pretend not to notice him. That's how my trip begins. My guilt trip, that is.

His eyes dart back and forth between me and the door with the forlorn look of a trust that has been breached. "I'll sit right here," say the eyes. "Right here until you get back. I won't eat. I won't wag my tail. I won't move from this spot until you return except to go outside and you know ... drain the old crankcase."

That's why I have a house sitter come in whenever I go away. A pet goes to a boarding kennel; Jake, on the other hand, has a friend do a sleepover and cater to his every whim.

I'm gone all of ten minutes when I double-back to pick up a

forgotten passport, and there's Jake, the loyal lad who was going to stand grief-stricken by the porch light until I returned, with his house sitter Carolyn sprawled out on the couch in the TV room, sharing a bag of Doritos and watching Disney's *Lady and the Tramp.* Jake, according to Carolyn, barked in the middle of some of the more suspenseful scenes, but otherwise was pretty well behaved.

So I leave unnoticed and not feeling so bad about it, except the very first tourist I run into coming off a tennis court at Loreto's Hotel Eden, halfway down Mexico's Baja Peninsula, is a dog named Thunder.

Sweaty and satisfied after two hours of hitting, Monica and I, the only players in the six-court complex, were heading back across the road to the hotel when we ran into a dog chasing a ball down the deserted street of this tiny Mexican village. We met him under the John McEnroe sign, which is falling off the front of this once-palatial and now-neglected sports center.

Thunder greeted us like long-lost friends and introduced us to

Chris and Trisha, good-natured people from Nelson, British Columbia, making an extended exploration of the Baja, hoping to set up an aviation business there. They brought their beloved brute of a yellow Lab along.

I thought it odd that they would have only one racquet and tennis ball between them, until they explained it wasn't theirs. It was Thunder's.

Like Jake, this beautifully fit and friendly mutt needs a lot of exercise and chasing a tennis ball in the cool of a moonlit desert evening is one of his favorite things to do. That and carrying out his civic duties as the self-appointed greeter of the village of Loreto.

I marveled silently at how dog people immediately relate to one another, openly and with some sort of dog-person warranty that you're trustworthy and decent human beings. Yes, subservient and somewhat soft in the head, but inherently good nonetheless.

Our first meeting involved only five minutes of conversation, some scratching behind the ears and a couple of sloppy dog kisses and then all of us were on our way five days later for a four-hour drive to the ancient city of La Paz.

This never happened when I was a cat person. Though a cat person's love for their pet is no less powerful, cat people tend to keep their relationships private and privileged. Cat people prefer a more intimate and confidential relationship. Cat people live their lives quietly and close to the vest. Cat people share books. Dog people share holidays.

Don't get me wrong, my cats were well looked after whenever I had to travel, but they didn't have agents working for them in far-off foreign countries.

Dogs, on the other hand, tell the whole damn world about their owners in one smile and a few quick wiggles of the bum. Whereas cat people are like Republicans or Conservatives, dog people are like Shriners on the three-drink-minimum plan. Every day's a pet parade, and your dog is always the Grand Marshal.

The minute I petted Thunder, I knew I was dealing with people who compared the nutritional benefits of dog food, walked and talked with animals and exchanged huge-vet-bill stories cheerfully. I suspected they also traveled with a dog bed in the back of the truck.

On the eve of the trip, over cold bottles of Corona at the simple house Chris and Trisha rented, we pored over maps for what can be a rugged and somewhat dangerous drive down the winding, coastal, paved path they call a road in the Baja.

Pit stops are few and far between so we broke the trip up into pee breaks. Not ours, of course—Thunder's.

Once on the road the next morning, and typical of guys, Chris was making such good time he couldn't bring himself to stop. I agreed entirely: No matter if you're hopelessly lost, in the eye of a hurricane or part of a convoy being evacuated after a nuclear strike—if you're a guy, and making good time, you don't stop until the vehicle begins to lurch. Then you continue as best you can on "fume fuel."

So with three of his passengers crossing their legs for the final hundred miles, Chris's only concern was for the dog in the back who travels in his own caged apartment and sleeps on a special futon bed.

Once in La Paz, Thunder immediately made himself the mascot of the El Moro Hotel. Walking routes were chosen so that the dog could romp in the ocean all the way into the center of town.

But suddenly along the route there was a nasty moment, a quick but telling confrontation between this beautifully groomed retriever and two mangy strays who passed us going the opposite way. They stopped, eyed Thunder with real hate that could come only from envy and bared their broken teeth at him.

Only the four adults around the dog kept the Lab from being attacked. The feral dogs dared not approach a person; this rule had obviously been kicked into them.

Thunder froze and watched them pass, surprised, innocent, confused.

But it was there in a flash, the great gap between the haves and the have nots, the same one all North Americans feel and pretend not to notice when they visit impoverished countries. It occurred at a lower level, but in exactly the same way. Soon, like all tourists, Thunder was distracted and galloping through the surf again as if nothing had happened.

Later, we selected an outdoor café that allowed the dog to join us on the street for lunch.

While we ate, most of our conversations began with something unique or cute or clever that Thunder had done, followed almost immediately by my rebuttal.

"Yeah, that's great, but let me tell you what Jake did when ..."

Then another Thunder tale, and I'd jump in with "That's nothing. Once when we were cross-country skiing and ..."

Until finally Chris and I had no other choice but to buy ourselves a pair of antique Mexican *pistolas* and shoot each other in the chest at twenty paces. (No, we did not do that.)

But I realized then and there that I had become a pathetic and uninteresting person.

If I had brought photos of my dog, I swear I'd have passed them around the table and probably showed them to the waiter.

When we got to comparing tick shampoos and heartworm pills, we all started to feel a little uncomfortable about how our lives had come to this sorry point.

Two certainties emerged from this four-person, two-dog friendship forged in a far-off land.

First, except in the throes of a natural disaster like snow and ice storms, Canadians are never quite as friendly and helpful to each other as they are when they meet in a foreign country. Suddenly the pride in being Canadian is without self-consciousness, and it's us against them from the get-go. People who wouldn't hang the national flag off their front porch are suddenly wearing the red maple leaf pinned to their bare skin when they're abroad.

Second, dogs dig themselves a hole in your heart so deep we might as well wear their photos on lapel buttons with the caption: "Slap me on the back of the head—I'm dog stupid!"

Jake may be at home watching canine chick flicks and sharing a bag of corn chips with his sitter, but he's still my travel agent and tour operator, deftly guiding my life thousands of miles away.

# CHAPTER SIXTEEN

# Singing the Dogcatcher's Blues

**YOU'RE A UNIFORMED OFFICER IN THE UNRULY,** quicksilver city of slow-driving tourists and fast-shuffling dealers where tourists from as far away as Japan come to watch water fall into a hole.

Your beat is the street, the mean street of strays that society has turned its back on, the mutts that don't amount to much.

Your name is Randy, you're an officer with the humane society in the city of Niagara Falls, and this is just one of a thousand stories in the Honeymoon Capital of the World, a true story set in the city that never sleeps because newlyweds keep everybody awake all night. It's the only place in the world where headboards come fitted with mufflers.

It's a dog-eat-dog world, and these are the dreaded dog days of summer, and nobody knows this better than you—you're the dogcatcher.

It's hot, and it's dank, and as you wipe the sweat from your face and neck with a handkerchief, the radio on your belt crackles your name and badge number.

They've found you again, and there's no dodging the dispatcher.

You jump into your truck, and you scream loud enough to part curtains all up and down the street. It's not failure or frustration, but there's that too. No, this time you just sat on that long-tooth, steel grooming brush you thought you lost weeks ago.

Your name is Randy, you're a hardworking officer doing an impossible job in an unforgiving city, and now you've got a sore bum to boot.

You've rescued eight dogs from as many steaming cars in the last two days alone, and this call is yet another emergency in four long years of emergencies in the business of life and death, chase and rescue, cats up trees and dogs crapping on the lawns of cranky neighbors.

This time it's red alert, and the crisis call comes from the parking lot at Marineland: An unresponsive dog locked in a van. Top priority. Expect the worse. Just do the best you can.

It's the same old story with slightly different details: medium-size dog in a red minivan with tinted windows and American license plates.

The darkly tinted windows are down a couple of inches and the Marineland security guys who are now helping you rock the vehicle have no idea how long the dog's been lying there, motionless on the back floor.

No response from the shaking, you go to Plan B, the water bottle. But still there's not a muscle moving from the heavy spritzing you're giving the dog.

The yelling, the hollering and the slapping of the sides of the van elicit the same response, which is no response, and you're down to your very last option.

No time to fetch a slim jim for a clean break-and-enter of the vehicle, you grab your trusty tire iron, and you're poised like a home-run hitter about to knock the passenger-side window into the

driver's seat. There will be serious damage for sure but maybe, just maybe, life might rise from the debris.

Your name is Randy, you're an officer of the peace about to commit a violent act on a foreign car with a possible corpse in the back. Funny, but your bum's not that sore anymore.

The window explodes in a thousand shards, and your hand finds the door lock even before all the glass hits the pavement.

Backup has arrived, and John, a man you'd trust with your life, rushes through the door, past the seats, and kneels beside the dog on the floor in the back of a van that's an airless oven on a sticky tarmac in a city that's hotter than hell.

A crowd has gathered around the van, and though the mood is silent, the moments pass loud with anticipation. They wait as you wait, with hope.

As John comes back out of the van, without the dog, he's white. He mumbles something as he passes you, and you catch only the swear word.

You dash in yourself, you crouch down, touch the dog, and feel for a pulse, and now it's you that's coming out of the vehicle, white, and silent except for the glass crunching below your shoes. You swear too, using different and even more profanity.

Someone in the crowd asks: "Is the dog dead?" You look at John, he looks at you, you both look down.

They ask again, and from somewhere comes the strength for you to respond: "It's stuffed."

"What???"

"The dog is a stuffed dog," you explain.

And there it is. You're an overworked lawman just doing your job, standing beside a car accident you created with one mighty swing of the tire iron on a torridly hot day in the city built beside a natural flush toilet. Your name is Mud.

At this point two things become quite clear to you. First, contrary to the TV ads, not everyone loves Marineland. And second, you can pretty much forget about ever being named Employee of the Month.

For one fleeting second you consider propping the dog up against the passenger door and putting the tire iron in his hands to make it look as if he tried to break out. But that would be wrong.

Besides, the glass is on the inside and there's too much of it to scoop up, and you're out of time anyway because the big American guy with the tattoos on his arms and the keys to the van in his hand is standing there glowering at you. Now his witchy wife is screaming about stupid Canadians and for-sure lawsuits, and the little girl is crying: "Mommy! Mommy! They're not going to take my doggie, are they?"

You haven't had a day like this since the rottweiler bit you in the ass, and the cat you were holding at the time, the one you'd rescued from the top of the telephone pole ... ran back up it.

Another call crackles on the radio, and you're out of there like Elvis on ice.

It's just another crisis, another critical call or complaint in the city that turns on the wheel of misfortune and spits out emergencies like quarters from a lucky slot machine over at the casino.

Your name is Randy, and your bum's not sore. But it's red.

# The Dog Rules

The dog's behavior outside in the yard should be
exemplary. You, after all, are judged by the conduct of
your dog, and the neighbors, though they deny
it, are always watching.

### RULE 1
The dog is not allowed to bark in the yard
unless he witnesses suspicious behavior.

### RULE 2
Snow, rain, wind and clouds do not
constitute suspicious behavior.

### RULE 3
The dog is to show the mail carrier
the same respect as you do.

### RULE 4
Okay, the dog can attack the postie if he or she is
delivering a really expensive brochure from the
government telling you what a great job they're doing.

### Rule 5
When a car pulls into the driveway, the dog must
sit to the side and wait until the car is stopped
and the engine has been turned off.

### Rule 6
Only then can he race to the nearest open
door, jump into the front seat and grin back
at you through the windshield.

### Rule 7
The dog is allowed to retrieve the daily paper
and drop it on the porch near the door.

### Rule 8
Okay, the dog can open the paper and read the sports
page, but if he does the crossword puzzle he has to
go back out and fetch the neighbor's paper.

### Rule 9
Okay, the dog can play in the neighbor's yard
with the neighbor's dog, but if he's going to stay for
dinner or sleep over, he's got to call home.

### Rule 10
And nobody cares what kind of cool blanket
the neighbor's dog got for Christmas:
"Your dog's not getting one! Okay?"

# *The Drive Up North—*

# *Yet Another Dog Disaster*

THERE'S NOTHING MORE INVIGORATING than a middle-of-the-night dash up north to cottage country.

You gas up and pack the car the night before. You go to bed early, with the last-minute list sitting on the kitchen counter reminding you to: Add ice to the cooler; place coffee and muffins on the dashboard; shut all windows in the house; make sure dog does his business; make sure you do your business; put water dish on back-seat floor. Once on the road you stop only at the gunpoint of an O.P.P. officer who shows you his badge as he is speeding along beside you.

As I've mentioned before, guys do not stop on car trips. Because when you get there, you want to be able to say to the first guy you meet: "Five hours! Waddijahave a flat? I made it in four hours and had to drive around several stalled cars with families who were stranded and waving white hankies at me!"

If we could do a deal with the air force in which a fuel plane gassed us up while we drove at full speed, don't you think we would? Then we'd never have to stop. Ever. We could live at the wheel of our cars with our dogs in the back seat using an onboard commode, visiting our families via the Internet and stopping only when we got to an ocean. Then we'd turn left. Oh sure, we'd slow down at the fast-food drive-thrus, but we would not stop.

So it's four o'clock on Friday morning, and if everything comes off just right, we can get past Toronto before it becomes a parking lot.

Cooler's filled, coffee's hot, everything's off and shut tightly, and my trusty companion Monica even shows up on time, which both pleases me and makes me a little suspicious.

As I put Jake's water dish in the back, I notice the side door is ajar. Which means the interior light was on all night. Which means Monica has to help me push the car down the driveway and out to the road, where she can maneuver her car into a position where her battery is on the same side as my battery, like nobody in a hundred years of designing automobiles ever thought to put the battery in the middle or make the jumper cables longer.

With the hatch back completely packed, I have to retrieve the jumper cables by going through the back seat from the inside of the car, and you know this had to happen ... I step in Jake's water dish.

Now we're a half hour behind schedule but ready to leave, as soon as I get my hands on Jake, who just bolted around the side of the house, probably chasing a squirrel.

Now here he comes wet and breathing hard with his ears and tail pointing down and YEOWWW!!! I guess in the dark a squirrel could look a little like a skunk, in the dark.

Jake had been drenched from head to haunches by nature's little can of Mace.

125

If there's something in the world that smells worse than a dog sprayed by a skunk, I'm sure Saddam Hussein would have it in his military arsenal by now.

So instead of clearing the Burlington Skyway at this point of the trip, I am on my knees beside the bathtub with a wooden brush giving Jake repeated washes and rinses with warm water, shampoo and—who keeps cans of tomato juice on hand for just such emergencies?—Extra Spicy Mott's Clamato Juice. If it burns a little, I think to myself, it's a bonus.

As the grayness of dawn begins to replace the black of night, I now have a dog that can't travel, a car I can't turn off, a big bottle of vodka in the kitchen cupboard that just lost any chance of ever becoming a Caesar and a soaker. Did I mention I love these family forays up north in the middle of the night?

With a bathroom strewn with towels I'll have to burn when I return, we set off with a guilt-ridden, stinking dog in the back seat, two edgy adults in the front and a car that's now overheating and running low on gas, and I think to myself, if only I had overslept, none of this would have happened.

So we stop for hot coffee at Tim Hortons, we average minus forty miles per hour getting through Toronto, stop again for fruit in Newmarket, gas up in Kleinburg, buy more and harder liquor in Barrie, and we arrive at our original destination, the town of Haliburton, the following Friday. And the dog still stinks.

But the neat part is we drove all the way with the air conditioner on to keep cool and the windows down to get rid of the smell. And the dog is as happy as a clam. It doesn't seem to matter to him that he smells like a bad one.

Actually we arrive by one o'clock, and once settled into our Willow Beach Cottage on Lake Kashagawigamog, the one that allows dogs, the first order of business is to tell a lie to the owner.

"As a matter of fact, Don, I did have a flat." (Sometimes it's just easier that way.)

The second order of business is a long walk along the trails near the golf course, a walk that is calmer than usual because all the critters Jake would normally chase and tree have smelled him coming from a long way off, and en masse, they've relocated to Gravenhurst.

The ducks next to the cottage weren't so wise.

And the screen door never stood a chance.

So my trip to the cottage is a series of small dog disasters. Not an earthquake, mind you, just a lot of little calamities that make sure you don't get too relaxed. Wait a minute—wasn't that the whole idea?

As a matter of fact, the earthquake hit in the evening.

I was busy in the small kitchen marinating steaks and brushing small potatoes with olive oil, when Monica came in to—here's a phrase that has a distinctly different meaning to different people—to help.

So I move toward the fridge and she's there, and I open a cupboard and hit her knee, and I go for the corkscrew in the drawer but she's now there and finally I suggested she'd be better off reading by the fireplace until dinner is ready.

And so she goes, in a slow smolder of rejection.

That's when she notices that although she's been banished from the kitchen, I obviously have no problem stepping over Jake and around Jake every time I move. Smack-dab in the middle of the action is where every mutt must sit. That, she shouted across the room, is apparently okay.

That's when I said something totally innocent that became the epicenter for the earthquake that followed.

That's when I said, as any dog lover would: "But honey, *he's* keeping me company!"

Never say that to a woman. By my estimation, it makes a long weekend way, way, way too long!

# CHAPTER EIGHTEEN

# *If Guys Were More Like Their Dogs*

DOGS, AS I'VE COME TO NOTICE, are approximately twice as happy as they have a right to be. For twenty-five years I enjoyed the company of cats—lovable, yes, but also surly, independent, suspicious and quirky. I loved it. It was like living with tiny New Yorkers.

Now, I'm living with MR. SOLID AND DEPENDABLE. I open my eyes first thing in the morning, and there he is, just inches beyond my nose, the big brown honest eyes, and the panting smiling face that says: "I love you, Bill. I mean it. You're like the greatest pleasure in my life ... since I've been neutered, that is."

Cheerful and ever-optimistic, with big white teeth glistening at you—every morning's like waking up with a motivational speaker.

I didn't name Jake, but if I'd had that opportunity Happy would have been a good choice, as in Happy Gilmour. Happy Hoover would cover his disposition and his food fixation.

That's why the first thing I do every morning when I wake up is pretend Jake scared the daylights out of me as he's lying next to me. I scream then leap out of bed and yell: "Who the hell are you? I don't own a dog! How'd you get in here anyway?"

And Jake's eyes narrow and his ears go on full alert, and he shows grave concern for all of a couple of seconds then the tail wags, the ears go back down and the smile returns to say: "You almost had me, ya big galoot. Seriously, I love you, man. Now let's go and get the paper, I gotta take a leak."

Watching Jake climb out of bed and immediately begin a series of stretches before he starts his day, it occurred to me that there's an awful lot of good habits we could learn from our dogs. Especially now that dogs are into brushing twice a day.

Think about it—beginning with the morning exercises and ending with a brisk walk around the yard before you retire—guys behaving more like dogs would not be a bad idea.

At best, our quality of life would be improved; at worst, it would be an awful lot of fun.

At least once a day I see Jake do something nutty like run around the house a couple of times for no reason at all, or plunge headlong into a hedge and use it as a back scratcher, or bark his fool head off just because he's happy, and I think to myself, where is it written that a guy can't do all the fun things his dog gets to do?

For instance, instead of always insisting on driving, a guy could significantly lower his level of stress by spontaneously jumping into a car, anybody's car, and going for a joyride with his head sticking out the passenger window. The wind in your face, you're drooling and yelling at other guys you pass—think about it. Who has time for road rage if you're mooning truckers all the time?

If guys were more like dogs, they'd be a lot more flexible because they'd always be stretching and rolling around on the ground and running from a rolled-up newspaper. And healthier too, because their meals would not consist of rich, fatty dishes, just a nice bowl of soy pellets with a helping of last night's spaghetti on top. And instead of two double martinis and a pound of pretzels, the happy hours would consist of wrestling on the living room carpet.

Likewise, a guy could have more fun if he got down on the kitchen floor on all fours and ate his supper out of a bowl. One bowl. No courses to prepare, no dishes to do, no table to set. He wouldn't get invited out to dinner very often but then again, he shouldn't because it's now his job to stay home and bark at shadows.

And instead of dialing 911, a guy biting an intruder in the ass would guarantee he'll probably not return anytime soon.

And think about it, if a guy ran to the door making excited noises and wiggling his bum every time a friend came to the house, a lot more friends would come to the house. If only to see it for themselves. We'd probably need crowd control if every time somebody knocked, Jake and I ran to the door on all fours, him with the pink bunny in his mouth and me with the bear-in-the-bag in my teeth.

And if someone gave a guy a cookie and a pat on the head every time he did the right thing, he'd make a lot less mistakes in life. Obedience school for guys wouldn't be such a bad thing, especially if the school had a football team and a pub on campus.

Look at all the extra time a guy would have for fun things like catching the Frisbee and chewing branches if he didn't have to shave and wash every day. And when it is time for a bath or a good brushing—somebody else does it for you! What guy doesn't like to be pampered and babied and dressed in a frilly little nightgown, talking baby talk

with a soother in his mouth and a pretty bow in his hair and ... Oh, sorry, I still can't get that Marv Albert thing out of my mind.

Where was I? Oh yeah, frequent dog naps would be a good thing for a guy as long as he remembered to hit the cruise control button on the dashboard before he nods off.

I think guys behaving like dogs would make social settings most interesting.

Just once it would be nice to see a guy dancing around the room wriggling his butt and howling when he wasn't drunk and desperately trying to keep up with the music.

Letting everybody fawn over you at a party would be great until you cold-nosed the hostess from behind and got yourself locked in the garage for the rest of the evening. Still, it's something the rest of the guests will be talking about for years.

Plus, at parties, you don't have to introduce yourself to strangers and babble on about who you are and what you do—you have a collar and an ID tag for everyone to see. And if you had to go after an obnoxious guest, at least he'd know ahead of time that you had all your shots.

Begging for food at the table may not be the kind of thing women find attractive in a guy, but it is the ultimate compliment to the chef.

And if a guy who stares deep into a woman's eyes while panting furiously could be satisfied with a simple bowl of cold water, it would save the woman a lot of time and trouble, including hauling out the big lie about having a headache again.

Now, nobody likes neutering, guys or dogs, but in some cases it's obviously necessary. I mean, if Hillary Clinton had neutered Bill rather than Buddy, instead of fetching a million dollars at auction, Monica Lewinsky's blue dress would just be way too small for her today. And let's face it, neutering the president of the United States would be far less expensive than impeachment.

If guys were more like dogs, their women would be much happier too. Suddenly they'd have everything they wanted: unlimited kisses, unconditional love, loyalty, a great listener and finally someone who understands the "relationship." For instance, a guy/dog would be perfect. They may bark and occasionally growl, but they'd never be able to get the last word in an argument. Plus, a guy on a leash is every woman's fantasy: "Sit! Speak! Down! Play dead!"

And there will be high humor, be sure of it.

A woman watching her man drink out of the toilet or relieve himself on the back tire of the neighbor's car—these are scrapbook moments that I believe even the grandchildren can someday appreciate.

And guys circling and sniffing new guys is way more entertaining than watching them shake hands and talk about the Leafs. Yes, humping legs could be a problem, but what guy doesn't occasionally need a good rap on the noggin?

Guys behaving like dogs—think about it—we could all go to the International Pet Show and get in for free.

Now, I may be toying with some silly man/dog role reversals here, but there's a woman from New York City who takes this very seriously.

Tired of picking up her man's dirty underwear and arm wrestling him for the remote, Karen Salmansohn wrote an obedience book for men. The title: *How to Make Your Man Behave in 21 Days or Less Using the Secrets of Professional Dog Trainers*.

Well, except for the petting part and frequent naps, I don't think most men are going to be too happy about being treated like a dog.

Salmansohn's theory is that you really can teach an old dog new tricks even if that particular old dog exhibits obstinate behavior as well as a little butt crack.

Some of Salmansohn's training tips are to keep the man's food bowl half filled so he's always wanting more, and therefore on his best behavior. Apparently a full water dish is okay.

She also sees stroking a man like a dog as a positive motivational technique.

And when it comes to discipline, she recommends the woman

134

always say: "No!" very clearly, so there's no mistaking the meaning. Like when he rolls over and wants his belly scratched during the motivational stroking session.

She also recommends the woman employ the old-fashioned reward system in the relationship. Find out what his favorite treats are and give him one for good behavior. "Make sure he begs a lot for it first," she suggests.

I'm guessing the sight of the man on his knees begging with his hands clasped together and his tongue hanging out doesn't really improve his behavior—she just gets off on it.

I say this because Karen Salmansohn's previous book was entitled *How to Succeed in Business Without a Penis*. I think it should be required reading for the man who agreed to go through her dog training course.

Salmansohn's next book is entitled *Whip Your Career Into Submission*. I can't tell you what it's about, but Marv Albert's on the waiting list for an autographed copy.

Half man/half dog might be a pretty strange idea, but as I look at Jake, who's just now returned from an evening with my neighbor Nelson—tired, with crumbs on his chin and beer on his breath—I think, hey, we're already halfway there.

# The Dog Rules

### RULE 1

All beer fridges would be refitted
with "bark-activated" doors.

### RULE 2

Now more than ever, we want the toilet seat left up.

### RULE 3

A ten-day cooling-off period would be required
before a woman could buy a broom.

### RULE 4

At least one piece of framed art in the house
must be *Dogs Playing Cards*.

### RULE 5

We must be allowed to get up on the vet's
examination table all by ourselves.

### Rule 6
And absolutely no thermometers!
We'll bark twice if we have a fever.

### Rule 7
In the game of fetch, we get to throw
the stick fifty percent of the time.

### Rule 8
All cable packages must include the
*All Dog All Sports All Day* network.

### Rule 9
And no fair nailing the doggie door shut
after we get used to running in and out of it.

### Rule 10
No way, you wanted us like dogs,
now it's your job to clean it up!

# CHAPTER NINETEEN

# Jake—The Stand-Up Comedian

# of Sunset Bay

DOGS ARE WHAT MAN WAS SUPPOSED TO BE—happy all the time.

Jake, for example, will occasionally surpass happiness and go right over the top to hilarity.

Slowly but surely and without my noticing, he has turned the whole neighborhood into his personal and very grateful audience.

Nobody drives up the driveway without Jake greeting them with the usual smile, bark and wiggle, and then he's into the first door that opens with "Okay, where's my treat?"

Even the courier drivers bring dog biscuits when they make a drop at my house.

For my neighbors, Jake has enacted the Milk-Bone Tax. To the east, Larry and Winnie, keep Milk-Bones in both their cottage and the garage. It saves time. They used to stock the small Milk-Bones until Nelson and Nancy, their neighbors to the east, started bringing the oversize Milk-Bones from Buffalo. So Jake, and this was not easy for a dog nicknamed Hoover, rejected their small offerings outright. He gave them looks of deep disappointment, dropped the biscuit on the deck and stared at them for a long time. He did everything but retrieve the box of jumbo Milk-Bones from Nelson's cottage and drop it at their feet.

To the west, Gaza, my Hungarian friend, brings Jake a hot dog every time he comes to his cottage, which in the summer is almost every day. On special occasions, the price he pays for the dog's friendship is a smoked sausage. And he better be quick to get out of his car when he pulls in, otherwise Jake will jump into the front seat with his bum resting on the horn as the poor man fumbles to unwrap the meaty treat.

Once he had established a fair and equitable toll system, receiving equal compensation for letting all my neighbors into their cottages, Jake moved on to "double dipping."

First, he hits Nelson up for a treat while he's still unloading his big black SUV for the weekend. Nelson keeps a box of Milk-Bones in the front seat for just such occasions. Grateful but greedy, Jake then whips around the side of the cottage, drops that Milk-Bone and proceeds to the lakeside porch, where he barks until Nancy comes to the door. He's got to move fast to preempt any communication between the two of them. Sure enough, Nancy gives him his "one a day" treat, and Jake returns with both bones in his mouth and, looking like a dog comedian impersonating a walrus.

Knowing I disapprove, Jake drops the booty on the patio and gives me that defiant, holier-than-thou look like: "I don't see you turning them down when they offer you a beer, Bill."

Then he takes his time eating both biscuits because he knows he can't repeat the routine until tomorrow. Except Nelson, who loves Jake dearly, pretends not to remember if he's already treated him or not. Meanwhile, my neighbors have no idea what they're in for this summer. Jake spent the past winter learning to open the doors of my house. Imagine if you will, a Customs and Excise agent who's adorable but no longer has to knock to come in. By August they'll all have locks on their kitchen cupboards.

Nelson is your classic dog lover. Jake loves him. Late one Sunday afternoon Nelson was mowing his lawn, and Jake was sitting next to the well watching him. Born to parents bred to work, Jake could watch others do it all day.

As I walked over to remind my dog where he lived, Nelson stopped the mower and came over to meet me. After a handshake and the weekly update of crossing over the Peace Bridge at Fort Erie with three inspection booths open to accommodate half a million Americans coming over on a Friday afternoon, Nelson paused and gave Jake a rare dirty look. It was then I noticed an oversize plastic wineglass sitting on Nelson's well cap with a bit of white wine left in the bottom.

"Dammit Jake," said Nelson. "That glass was full when I put it down."

Jake gave me a nervous look then licked his chops.

"So Bill," said Nelson, draining the last of the wine from the green goblet, "how was your week?"

Now that's a man who loves dogs.

And he has more than one reason not to. Over the years Monica has taught Jake to bring things to people.

So well has Jake taken to this task that at Christmas, Jake plays a pretty good Santa Claus.

First Monica works him once around the room at the family home in Welland, introducing Jake to everyone in the room. Then she gives Jake the gift and the name on the gift.

"Bring to Paula! Bring to Tommy! Bring to David! Bring to Sue!"

And like a postie on an indoor route, Jake delivers on cue.

Plus, when "Bring to Bill!" comes around, he gets to open the gift too.

On warm weekend nights along the lake in Wainfleet, cottagers sit around barrel fires on their breakwalls laughing and lifting glasses. It's a fine tradition. From my vantage point, I can count up to a dozen such fires burning around the bay from Morgan's Point to Rathfon Inn on a typical Saturday night in the summer.

The fires are spellbinding, luring you into a drowsy trance while the shadows that change with the intensity of the blaze dance all around you like black ghosts.

Nelson was in just such a mellow mood, sitting by himself in a lawn chair beside his fire with a nightcap in his hand when—and we cannot figure out why she did it—Monica put a flashlight in Jake's mouth and said: "Bring to Nelson!"

The flashlight was on, and Jake was off!

Jake ran straight for Nelson, and then past him to something more interesting, more promising of food farther down the beach.

The next thing I heard was the loud, panic-stricken voice of a

141

man in peril.

"Holy %@#*! What the hell was that?" yelled Nelson.

A black dog in the dark of night is impossible to see. But a fast-moving beam of light, apparently attached to nothing, is hard to miss.

"Nancy!" screamed Nelson toward the cottage. "What the hell was that?"

"What was what?" came the reply.

"A light! Some kind of light just went across the yard!"

And just then Jake came speeding back toward us.

"Geez, there it goes again!" yelled Nelson, and this time he got up and went into his cottage.

And none too soon. Another minute, and two people on my side of the property line might have died violently from laughing internally.

Jake loved it. He tried to get the flashlight back to scare the hell out of everybody relaxing on Sunset Bay that night.

I don't know if I've ever seen a man so relieved as Nelson the next morning when I apologized and explained.

Nelson's mind had run the gamut of possible explanations, from a UFO to a hallucination, from a sign from God, to the world's biggest firefly.

"I'll tell you the truth, Bill," said Nelson in a confessional tone, "I damn near quit drinking right there and then!"

And that's when I realized how a harmless prank can sometimes do a lot of damage. Because a man on his breakwall by a beach fire without a drink ... is like a dog on his blanket without a bone. It's just unnatural is what it is.

Although Jake is still the stand-up comic of Sunset Bay, the flashlight bit is no longer in his bag of tricks. That's just way too dangerous a routine.

# The Dog Rules

AS THEY APPLY TO STAND-UP COMEDY

### RULE 1
Don't start with "Good evening, ladies and dog germs."

### RULE 2
No foul language and stay away from toilet humor.

### RULE 3
Okay, you can do that joke about stopping at the
toilet for a quick drink after work. But that's it.

### RULE 4
Remember, licking yourself is funny only to female dogs.

### RULE 5
Do that bit about the bottle mix-up and
getting shampooed with Nair. That's funny.

### Rule 6

You can't do better than that: "Me The Pooper/He The Scooper" routine.

### Rule 7

And when you said to the vet doing the examination: "Rectum? Damn near killed him!" That'll work.

### Rule 8

No. That "Do you do it while your pet's in the room?" thing has been done to death.

### Rule 9

Be careful with the "I really love cats ... they taste like chicken" bit. They're organized and looking for an inciting incident.

### Rule 10

"Come on. Get over yourself. We're not really happy to see you come home. We're just buying time before you find the garbage on the kitchen floor." That's a closer.

# Walking the Dog—

# The Great Two-Way Tinkle Tour

*Psychiatrist: "So how long have you believed that you're a dog?"*
*Patient: "Ever since I was a puppy."*

YES, THEY WILL DEFINITELY DRIVE YOU NUTS THESE DOGS.

Walking your dog, for instance, is supposed to be a healthy and pleasurable outing. Good for you, good for him, something your doctor and his vet can shake hands on.

And the collar-and-leash system was invented to ensure the maximum amount of safety and discipline on just such a neighborhood hike. (And let's be very careful about using the word "hike" around a dog who's been inside all day and paying an awful lot of attention to that rubber plant in the corner.)

Take the case of the rambunctious, full-grown but still immature male. This collar-and-leash control system can keep him from tear-assing across neighbors' yards, running wild in the streets, soiling private property and harassing every female he meets. If he flatly refuses to wear a collar or be tethered to a leash, might I suggest marriage counseling?

But enough about your personal problems. I'm here to talk about dogs.

As the owner, provider and guardian of a dog, you have certain rules to follow:

When walking, your dog should be on a good strong, short leash. Remember, the length of the leash is directly proportional to the amount of trouble your dog can reach. Long leashes are fine for offroad walks in fields, parks and along shorelines, but in confined areas they should not become jump ropes for older people carrying groceries.

And, of course, you have an obligation to scoop or bag everything he leaves behind. Let's keep in mind that the great outdoors is not a place for pet poop. It's a place where people can enjoy fresh air and lush scenery while they empty their car ashtrays and turn the landscape into a natural dumpster with fast-food garbage and pop cans. Compared to dogs, people are pigs. (Sorry, but I'm one of those people who spends the occasional Saturday morning picking up other people's garbage along the lakeside, the person you drive by and say: "Get a life!" And in an amazing reversal of roles, Jake's the one holding the bag.)

There are certain common-sense rules that apply to walking. However, since dogs began domesticating humans back in the late fifties, they have their own agenda. All dogs, as you may have noticed, have a slightly different take on the rules that we laid out to govern their lives.

For instance, the parameters that constitute a walk for a dog, as

147

outlined by a dog and for a dog are based on one central recurring theme: If it moves, chase it, and if it doesn't move, piss on it.

This behavior is irrevocably instinctive and I might add, the very same military strategy used by such successful conquerors as Napoleon, Hannibal and Warren the Whizzer.

This code of conduct, when applied to your daily dog constitution, is both constant and pervasive. In other words, everything you encounter on your walk, from the time you leave home to the

moment you return, is either running for its life or getting damp. A walk with a dog is not a stroll but a fresh-air peeathon.

"Tree it or pee it!" would be a good slogan if dogs wore T-shirts instead of pullover sweaters and yellow rain slickers. (Please, can we stop dressing up dogs? Cats are keeling over with laughter.)

Try never to say the word "walk" until you're dressed with leash in hand, plastic bags in your pocket and ready to go. Verbalizing the word "walk" even in casual, unrelated conversation that the dog happens to overhear is a promise etched in stone. And that stone needs a little sprinkle, because when you say "walk"—it's show time!

In fact, it's always best to spell the word "walk." Unless you own a Border collie; then spell it backward.

Your dog should walk in such a manner that the leash remains slack. Tight is all right. But if your dog is exerting more than 200 kilo-kanines per human stride, then that's not walking. That's water-skiing on asphalt.

Squirrels are the natural prey of all domesticated dogs. Back when dogs were wolves, their quarry was deer and moose, which shows you how low their expectations have dropped in just a couple hundred years.

Squirrels believe their purpose on earth is to torment your dog by running to the nearest tree and past it to the farthest tree; then back to the tree in the middle and up and over the branches, jumping to the power line like the last of the Flying Wallendas until your dog is four inches from the telephone pole and still motoring at full throttle. If you can't train your dog to stop chasing squirrels, at least try to get him to wear a helmet.

A long walk with the average dog is like a total all-inclusive tinkle tour.

Hiking across country, in dog terms, does not involve a backpack. It involves the right hind leg. The perils of peeing too much are just a rumor in the dog world. Biologists and zoologists alike are at a loss as to how a dog can pace himself to the extent he is able to urinate on every single inanimate object he encounters on a five-mile walk. Equally befuddling is how they still have enough left over to give your car tire a final spritz when you get back home. A double-bladdered, two-humped camel that just inhaled an artesian well couldn't match this feat.

And not only does your dog have to cover the scent of the previous dog, he has to pee higher and ... we're not sure what this means in human terms ... better. There's a hike-the-hind-leg hierarchy at work here. Those who can, pee high, pee proud. Those who can't, often suffer from what I imagine to be the short dog syndrome. Show me a small dog that wants to fight a big dog, and I'll show you a tiny tinkler.

And when your dog meets another dog on the walk, it's an encounter known as "one of life's most embarrassing moments." Keeping a straight face while conducting small talk with the other dog's owner as the two of them engage in serious nose probing—this should warrant some sort of acting award.

It's reddening-from-the-neck-up time. Blushing, you look away, point to something in the opposite direction which doesn't exist, say something unintelligible to the other pet's owner and nearly sprain an eye surreptitiously sneaking a peek to see if they're still at it.

Please, relax. Do not feel ashamed. Thanks to the Internet, your kids are probably watching humans do much ruder things while you're out on this walk with your dog.

This cursory inspection takes place for very good reasons.

The first general, broad-based sniff and nudge is to determine size, what was for lunch, excessive perspiration and vitality of coat.

I swear I've seen Jake look back at me during one of these malodorous examinations of a gorgeous golden retriever with a smile on his face that said: "Oh, Bill, it's not the shampoo. It's the conditioner!"

Then there's the more personal inspection of those regions best left to veterinarians wearing gloves. This is where dogs get down to the real business of sniffing, and as any mutt will tell you, there's no business like nose business.

While nose probing, a dog is, in short order, determining such things as gender, habits of hygiene and sexual readiness. (I'm told that sometimes evidence of carpet burns on a dog's bum will cause the other dog to roll around on his back howling with laughter.)

Basically they do everything but ask the other dog to cough.

And the first one to walk away from one of these sniffathons wins. The first dog to walk away from the big public whiff is saying, I smell better than you do otherwise you wouldn't still be hanging around. I mean, this is not the kind of tournament in which you want to come second.

The obvious question for pet owners who are mortified by these raw and spontaneous public demonstrations by their dogs is: Should we put clothes on our animals, especially pants and underwear?

No. No species of dog, except perhaps the boxer, is going to be comfortable wearing shorts.

The overriding rule here is: If God wanted dogs to be modest, He'd have given your pooch a pouch.

It's important to remember that until humans began smuggling illegal drugs and weapons through airports and across international borders, dog sniffing was not considered a professional job.

My question is: how come we have to buy smoke detectors? If a dog can be trained to smell an ounce of cocaine buried in an airtight piece of Samsonite luggage, how difficult would it be to get him to bark when the house starts to fill with smoke? And if we can train a dog to "Sit!" and "Speak!" and "Roll over!"—how come we don't teach them to shake paws when they meet other dogs instead of that sniffing thing?

However, the next time you must stand by and watch your dog commonly assaulting another canine with his nose, do not be embarrassed. Do not scold the dog. Do not tie the leash in a noose around your neck and climb the nearest tree.

Remember, it is a vital and instinctive information-gathering session.

And let's not be too harsh here. The spontaneous sexual inspection and random mounting is not limited to dogs in parks. See the next episode of *The Jerry Springer Show*—"Young Women Who Seduced Their Uncles and the Boyfriends Who Married Them (the Uncles)."

# The Dog Rules

A successful motor holiday involving a dog depends
on careful preparation and the strict adherence
to a few basic tips.

### Rule 1

When packing the car, allow enough space in the
back seat for the dog to sit upright without bumping
his headphones on the roof.

### Rule 2

Once the dog's in the back seat, lock
the picnic basket in the trunk.

### Rule 3

Before leaving, explain very loudly:
"Not stick! Emergency brake! Not stick!"

### Rule 4

Games of tickling and goosing may involve
the dog, but not the driver.

### RULE 5

In the game of "Do you see what I see?"
it's unfair to use color descriptions with dogs.

### RULE 6

Keep the dog's identification card and health certificate tucked
in the brim of his Tilley hat for the duration of the trip.

### RULE 7

At least thirty-five percent of the dog must be
inside a moving vehicle at all times.

### RULE 8

When the dog puts his paws over his ears, that's enough of
"How Much is That Doggy in the Window?"

### RULE 9

Having the dog wear a variety of hats on the trip, like
those of firefighters, police officers, construction
workers and cowboys helps reduce road rage.

### RULE 10

No postcards mailed from your dog to neighborhood
dogs. Dogs touch, they don't keep in touch.

# CHAPTER TWENTY-ONE

# *Jake in the City*

I'M NOT MUCH OF A CITY PERSON, and neither is my dog.

Jake will, however, tag along, reluctantly, if I have business that occasionally keeps me overnight in Toronto.

Jake stares in amazement at the skyscrapers, looks at every passerby like he should know them and tries to jump into every open car door. And he has great difficulty doing his business on anything but a patch of grass. So essentially, he behaves the way everybody from Wainfleet does when they're excited and in the big city.

Jake's favorite hotel is the Marriott near Pearson International Airport where I can now get him into the room without putting on sunglasses and walking into all the furniture in the lobby. It's completely gone to the dogs (see The Dog Dictionary).

Jake has stayed at the downtown Holiday Inn in Toronto, where he made an appearance on *Basic Black* with host Arthur Black. It was a special live edition of the show with a sold-out audience at the CBC's Glenn Gould Studio, and by the time we left, Jake had met every member of the sold-out audience personally. While I was trying to get him to do a trick on national radio, Jake was running up and down the aisles pretending to kiss and hug everybody while frisking them for biscuits.

Suffering from poor billing, Jake had to follow a couple who chased tornados for a living and Backwards Bob, a guy whose profession is talking backward phonetically as well as naming the capital city of any country in the world, along with its population and size in square miles. Jake did not like to be called Ekaj and he seemed stunned to learn that Iceland had only 222,000 people spread over 39,768 square miles.

Also, because it was a radio show, he was most upset that only people in the studio audience got to see what a ruggedly handsome brute he is.

Since any dog can bark for a treat, Jake was there to perform his one and only trick. On cue, when I say, "Speak softly," Jake will speak so softly you almost can't hear it. It's a whisper bark. And it's so cute, people go: "Awwww."

So, live on Arthur Black's CBC Radio show, with treats in hand, I told Jake to "Speak softly." And he barked as loud as he could. Not thinking it was much of a trick, the audience booed. We left, and I haven't been on the show since. Honest.

Jake has even been to a Major League baseball game and watched the whole thing stretched out in bed.

When I was in Toronto to read at the *Word on the Street* festival, my publisher put us up at the SkyDome Hotel.

It's important to note that Jake was allowed in the hotel in the SkyDome but not in the ballpark itself. Professional baseball has a very strict "No Pets Allowed" policy. I know because I actually interviewed the pioneering mutt that broke the ban on dogs in baseball. It's not exactly the Jackie Robinson story of breaking the color barrier in baseball, but it's true.

In the summer of '93 I wrote and coproduced a baseball movie entitled *Chasing the Dream*, a story about three young professional ballplayers struggling in the New York Penn League, trying to climb the minor league ladder all the way up to the majors.

On the film shoot, traveling on a bus with twenty talented, fresh-faced kids with laser arms and raging hormones, I was the "geezer guy." In a half-dozen ballparks spread across the state of New York, my job was to uncover the aging and colorful characters of baseball's past. I interviewed the old guy who remembered Babe Ruth barnstorming through these towns in the thirties and the eighty-year-old woman who remembered kissing Mickey Mantle in Oneonta on Mickey Mantle Day. All this we would use to spice up the game play on the field, which gets tedious after a while.

At the park where the Glens Falls Redbirds played, apparently there was no color.

The manager had nothing for me—no octogenarian fan who remembered Ty Cobb, no wacko who stripped to his shorts to the beat of the national anthem, no local barber who gave lucky ticket holders haircuts at home plate.

But as I was leaving his office to bring the bad news to the director, the manager said: "Oh, we got a dog."

"A dog?" I asked. "On the team?"

"No, we got a dog who owns a season ticket. That's him behind the backstop."

I was there in a flash with a cameraman, filming this mangy, overgrown French poodle as he barked at the batters and chased foul balls into the parking lot.

Dutch, the dog's owner, told me the story on camera, how they loved the dog and they loved baseball, so they sat down with the management of the ballpark and worked out a deal.

As I recall, the dog's name was Prince. Or maybe it was the dog formerly known as Prince. I forget.

(I'll never forget Dutch's wife, Biddy, because he affectionately referred to her as "the old Biddy." In small-town America, that's funny.)

159

So Dutch told me the story of how they bought the dog a season ticket to the Redbirds, and how he's now the only season-ticket-holding dog in all of professional baseball. And then Dutch started wandering off topic, so I tapped the cameraman on the knee, the signal to wrap it up.

And here's why, with anything involving a dog, nothing ever comes off exactly as you hoped for.

Because, as soon as that camera shut down, Dutch looked from side to side like this was on the Q.T., and he leaned in and said in a very low voice: "Oh, and by the way, Bill ... when we negotiated to buy Prince the season ticket ... we worked it out in dog years so's we got him the senior citizen's discount too!"

I turned on the cameraman: "Why'd you turn that thing off?"

"You told me to," he said defensively.

Dutch would not retell the story on camera for fear they'd charge him full fare next year.

And it was lost: One of the greatest moments of humor in baseball history involving a dog four rows behind home plate.

Meanwhile back at the SkyDome, where Jake was not allowed to sit in the ballpark, he was cleared to stay in the hotel.

Immediately after checking in, things got off to a shaky start when, for the first time apparently in his life, Jake stepped onto a floor that moved. I think there's something instinctive in a dog that makes him think when he's on an elevator he's actually in an earthquake. Moving floors, see-through steps and grated bridges make Jake a cautious walker. Instantly his legs spread wide to maintain his balance on the elevator, and a look of grave concern swept over his face.

We got into the room at precisely 1:10 p.m., five minutes after the game between the Toronto Blue Jays and the Detroit Tigers started, and I had barely set the luggage on the bed when the fireworks started.

I mean it. I hadn't even got the door shut when second baseman Craig Grebeck, one of the weakest hitters in the Jay's lineup slammed a home run over the right-field fence, and the ceremonial fireworks went off directly above our room. The noise was so deafening it sounded like the explosives had gone off *in* our room.

I lunged for the door, but it was too late. Jake, who's terrified of thunderclaps, gunshots and fireworks bolted like a bat out of hell down the hall. So, as Grebeck is circling the bases, I'm sprinting through the circular corridors of the SkyDome Hotel screaming for my dog to come back.

He's nowhere in sight, and I imagine myself calling the front desk to ask if Jake Thomas has recently checked out of the hotel ... at approximately 120 miles an hour.

If he made it to the field, we were really in trouble. Jake would have no problem catching a pop fly or even a hard-hit grounder, but the players on both teams would be at least an hour getting the ball away from him.

On the other hand, with Roger Clemens on the mound, Jake could add a very wet spitter to the baseball star's pitch selection.

Help! Who do I know at the SkyDome? Paul Beeston's a good friend of mine, but he's president and CEO, so I'm guessing he's got better things to do than pursue a dog ... although I do recall he was responsible for signing pitcher Mark Lemangello back in 1979.

The only player I know personally is Shannon Stewart, the only kid from our minor league film to ever make it to the majors. Now the speedy and hard-hitting center fielder of the Toronto Blue Jays, Shannon would at this moment be standing at home plate waiting to high-five Grebeck. He's busy.

The only ex-player in the building I know is George Bell. I wondered if he would help me, and then I thought, who am I kidding? In ten years of playing outfield for the Toronto Blue Jays, George Bell couldn't catch anything.

Thank God for women. Jake adores them.

There he was, waiting at the elevator doors, charming the ball hat off a woman who was down on one knee shaking his paw and calming him down.

In the event of another home run, I moved all obstacles out of the path of his escape route to the bathroom and used a blanket to rig a dark cell below the sink in which he could hide.

So that's how our adventure to the big city started, a desperate footrace around the hotel, followed by a touch therapy session under the bathroom sink, with me trying to convince him it was not some kind of cruel practical joke. I did not bring him here to watch fireworks.

Then we settled down on the carpet in front of the floor-to-ceiling window, and Jake watched his very first Major League baseball game.

Much like the 40,268 fans on this, the last day of the season with the Jays nowhere near a playoff spot, Jake wasn't all that impressed either.

He made this strange throaty noise every time baseball's bad boy Jose Canseco came up to bat, but then so did all the University of Toronto engineering students sitting way up in the cheap seats.

He loved the popcorn and cold beer, especially when I was out of the room.

Jake was a little nervous with just a single pane of glass separating him and a hundred-meter drop down to center field.

I loved it. Finally I had the exact perspective of watching a Major League baseball game and staring out at 40,000 people as that couple who were caught having sex up against the window of the room next to ours a year earlier. They must have thought they were in the old ballpark, the one named Exhibition Stadium. Boy, you have to be pretty confident to make love in front of 40,000 people, every one a critic.

The game went on for well over four hours. Jake got so bored, he watched the last four innings propped up in bed. He napped through five choruses of the crowd singing "Take Me out To the Ballgame."

Roger Clemens struck out eleven batters, and Shannon Stewart singled in the winning run in the thirteenth inning, just as I had the phone in my hand, about to call the manager. I mean, four hours with intermittent fireworks—Please!—keep the noise down, my dog is trying to sleep.

# The Dog Rules

AS THEY APPLY TO THE BALLPARK

No, they're not kidding.
"No Pets Allowed in the Ballpark."

### RULE 1

Except for dogs in possession of a valid season
ticket with their full name and signature.

### RULE 2

No, puppies do not qualify for
"Kids Under 12 Get in Free."

### RULE 3

No chasing mascots unless they're cats,
squirrels or fat birds that can't fly.

### RULE 4

No begging. That's the job of the sports agent.

### RULE 5

Dogs keeping score by barking can be
annoying to other fans in their section.

### RULE 6
No racing out and retrieving the ball even if three guys
on the home team have let it get through their legs.

### RULE 7
A dog is not allowed in the outfield no
matter how bad he has to go.

### RULE 8
Absolutely no biting the players unless it's John Rocker
and then only if the dog has all his shots.

### RULE 9
A dog is never to steal second base without getting
the green light from the first-base coach.

### RULE 10
After your dog wins the Frisbee-Catching Contest
do not offer to loan him to the center fielder.
That's bad form.

CHAPTER TWENTY-TWO

# The Difference between

# Cats and Dogs

TO THIS POINT IN MY LIFE I have lived with both cats and dogs, six felines and three dogs, to be exact, and I've noticed that they are exceedingly different.

Generally, cats are smaller and more cynical. Dogs are bigger and more naive.

Cats, both male and female, are a lot like women. Similarly, all dogs are like men.

Cats are subtle. Dogs are "Duh!"

A cat is a clever little creature who can often make a guy look stupid. A dog, however, can flatter a man by embracing and rejoicing in his stupidity.

Like at midnight if the fridge door hits you in the arm, knocking the plate of leftovers and the quart of milk crashing to the floor, the dog doesn't try to figure out what or why it happened. He's immediately and enthusiastically involved in the cleanup. A cat will sit and analyze the situation: looking at you, looking at the mess, looking eminently superior. Despite what you may have heard, a cat will in fact worry over spilt milk.

If you ask a cat: "Jawannagoferawalk?," you get nothing in return.

Say "Jawannagoferawalk?" and the cat will burrow even deeper into the pillow of the couch, but before he nods off he'll give you one weird last look like you need a thorough psychological assessment, the kind that involves a whole team of doctors with thick, round glasses.

But if you ask a dog: "Jawannagoferawalk?" the dog exhibits behavior similar to that of Richard Simmons with a colony of ants in his pants. He's wild with exuberance. He's animated. He's with the program.

You couldn't get a dog more excited by saying "Jawannagoferawalk?" or "Jawannagoferaride?" if you got down on the living room rug and wrestled with him while wearing a garland of panfried pork chops.

Dogs will follow you into a ground war in a Pacific jungle if your request begins with "Jawanna." On the other hand, it takes a week for the scratches on your arms to heal after getting your cat to the vet's.

Dogs are like drunken frat brothers, followers, joiners, good-time Charlies. Cats are like Howard Hughes. Spooky, twisted, easily made irritable.

If you watch television with a dog and turn it up loud when Eddie on Frasier starts barking, a dog will get excited and bark back.

If you watch television with a cat and turn it up loud when Garfield is on the screen, the cat will turn down the volume, then switch the channel to PBS's *Masterpiece Theatre*.

Dogs are like Shriners in a parade: fun-loving, go-cart driving, impossible to embarrass. Cats are like ambulance drivers at rock concerts: Very, very weary.

If you're reading or writing, a dog will take that opportunity to sleep at your feet. If you're reading or writing, a cat will take that opportunity to get between you and the page.

Cat heirs, like the dozen or so owned by Mae Lunn of Halifax, Nova Scotia, who died last year and left them $317,000, know a good thing when they see it.

Those cats will first seduce and then supervise the live-in caretaker appointed by the executors of the will and live out their lives napping and nibbling in the manner to which they'd become accustomed, just as the court order called for.

A dog would take the $317,000 and go to the track.

Not the dog track, mind you. Oh no, they're absolutely opposed to the exploitation of their brothers in the inhumane, slavelike sport of dog racing. But horses? Oh yeah, dogs love to watch big, stupid ponies run around in circles for money.

Cats are fastidious, worrisome and usually skinny. Cats are like Calista Flockhart. Just try giving a cat a pill. It's why should I take the pill, what's in the pill, will the pill cause such sudden weight gain that when I stand sideways I'll actually be visible. Giving a cat a pill is like a day-long horking marathon in which the whole bottle eventually dissolves into a big ball of spit in your hands. You could molecularly bind that pill with beluga caviar, and it's still coming out of that cat's mouth like a mint from a Pez dispenser.

Anybody who has ever tried to get a cat to swallow a pill winds up wondering if it would work at the other end.

Dogs, on the other hand, are a lot like Norm from *Cheers*.

Give the dog a pill that's rubbed in anything resembling food, and he has one comment on the whole procedure: "Gulp." And then the sit-and-wiggle that says: "Gimme another one, Bill. I've been a good boy today."

Tell a dog you're going to the corner store, and he'll run through a plate-glass window to go with you. Tell a cat you're going to the corner store, and she'll run through a list of things she'd like you to pick up.

A dog protecting his territory will bark, growl and, if necessary, bite. A cat can sit on the other side of the room and stare at you until you confess something you're not even sure you did.

Dogs are devoted. Cats are cool. Dogs will lie beside a wounded hunter all night to keep him warm. The first thing a cat will do is wipe his paw prints off the gun.

Dogs are nosy. Cats are fanatical. A dog will nose open the door of the pantry to see if his favorite food is there. A cat will jump into the pantry and rearrange the tins of food, putting the dog food way in the back.

Dogs are cordial. Cats are aloof. A dog will rush up to visitors, offering kisses or his favorite toy. A cat will sit on the chair near the door and interview them with her eyes. Dogs make lousy guard dogs. Cats would make terrific personal managers.

A dog will run downstairs in a thunderstorm and hide in the back of his cage. A cat would run downstairs in a thunderstorm only to flip the lock on the cage and leave that dog in there for days.

Dogs will beg for food. Cats will agree to attend a tasting where—and if this is not the theme, they're a no-show—presentation is everything.

A dog is afraid of the vacuum cleaner. A cat is afraid she's not getting deep enough into the nap.

Dogs love to pose, especially for family photos. Cats like to hide behind the couch, especially during family photos.

Dogs can stimulate behavior in people. When they yawn, we yawn. Cats also stimulate behavior in people. When they yawn, we try to come up with more interesting conversation.

With a dog, when the luggage comes out, depression sets in. With a cat, when the luggage comes out, the cat gets in.

A dog will jump between two people having a domestic argument. A cat will call 911.

A dog will roll onto his back to get his belly scratched. A cat will roll onto her back because it's an excellent defensive position.

For a cat, cleanliness is next to godliness. For a dog, cleanliness is next to last on his to-do list. A cat is a completely self-cleaning unit. A dog is like a walking lint brush, attracting loose debris. For him, a shake is as good as a shower.

A dog having a warm bath is the picture of indulgence and contentment. A cat having a warm bath is like Mia Farrow's childbirth scene in *Rosemary's Baby*. A lion tamer who works with six big cats in a cage would not give a cat a bath on his own.

A dog's attitude around the house is "Don't Worry, Be Happy." A cat's attitude around the house is hate everything immediately, and better options will be offered.

Talk nice to a strange dog, and he'll sit and offer you a paw. It's the start of a friendship. Talk nice to a strange cat, and she'll turn and give you the finger. It's the end of the discussion.

A toad on the patio is something a dog will bark at and chase. A toad on the patio is something a cat will torture until the toad wishes he'd encountered the dog instead.

Dogs sometimes have nightmares where they yelp and twitch like they're being attacked by an evil force. Cats watch with delight and like to think they are that evil force.

Dogs will serve as toys for kids. Cats see children as their equals.

A dog will vacate his owner's favorite chair, but ever so reluctantly. A cat will vacate her owner's favorite chair immediately ... so she can jump back up into the lap.

Bottom line: Over thousands of years the canine has emerged to become man's best friend. Over the same period of time, the feline has emerged to become man's best fiend.

Remember the one and only Household Pet Rule: To a dog, you're family. To a cat, you're staff.

# The Dog Rules

With some exceptions, dogs and cats
have a healthy mistrust of each other ...
Let's keep it that way.

### RULE 1

No interbreeding between dogs and cats. The world
is not ready for an animal that chases squirrels ... all
the way to the top of the tree.

### RULE 2

If leash laws for cats do come into effect,
dogs are not allowed to be walkers.

### RULE 3

Barking, okay. Yeowling, all right. But no harmonizing.

### RULE 4

Cats are not allowed to put the dog's food in the clothes dryer
and slam the door once he's in there. That's not funny.

### RULE 5

Cheer up, it's not your fault—even cat psychologists
don't fully understand their mood swings.

### RULE 6

I know, it seems like an awfully dramatic scene to produce
one little hairball, but that's just the way they are.

### RULE 7

No, actually I've never heard of a "working cat."
You've got a point there.

### RULE 8

Sure a dog can knock the receiver off the cradle when the
phone rings, but will he take a message for the cat?

### RULE 9

No, the beach is a theme park not a litterbox. One
more time and I'll cover "you" up with sand.

### RULE 10

No, that cat's not dead. She's hunting.

# CHAPTER TWENTY-THREE

# *Canine Kodak Moments*

**ALL MY LIFE I HONESTLY HAVE NEVER SEEN A DOG SMILE.**

When I get home and walk through the door, Jake drops his head, a little embarrassed, and then breaks into a big, toothy, demented smile pronounced "schhhmile" at my house. It's amazing, and it never fails to make me laugh.

But when I get home from a vacation or a business trip, there is no smile. There is unbridled, nervous excitement that threatens to send a wiggling bum right off the body and smashing against a far wall. And there is definitely snorgelling. This is a variety of guttural noises emitted during a long hard hug that come from somewhere down deep in the soul that say: "Don't ever do that again—I almost gave up on you that time."

Every day, except when it's frozen over, I throw sticks far out into the oncoming waves of Lake Erie for Jake to fetch. I do this because he loves the game, the exercise is good for him and I love to watch the burst of speed and the splashy collision when he hits the water in full stride. And he doesn't just assault any wave, it's the white curl at the top that he attacks. Then there's the barking through bark and

the growling over wood, but the best is the seven-second shake.

The shake begins with the head, followed by his chest and torso, with his bum coming late to the party. When he stops, his head quits first, but his bum continues like it's got a gyrating little life of its own. Jake looks straight at me, his head perfectly still and tilted in curiosity as to what I'm laughing at—and that bum is still wiggling from side to side, spraying water for a full three seconds after the main furry flourish is over. It's like it's on some sort of delay system, set off by a weird coin toss: heads I quit, but tail, you keep on going. I love the shake.

Jake loves to have a sweatshirt or blanket or towel draped over his head. Instead of tossing it off, he likes to walk around with it covering his head, bumping into things. But once it does fall off, he takes it in his teeth and tries shaking it to shreds. He snaps the towel from side to side, galloping and growling, and rolling over and over on it. Then it's back on his head like a loose-fitting hood, and he stalks people like a deranged clown.

Jake has carefully categorized our guests. Ordinary people, friends he sees frequently, get the bear when they show up at the door, the ragged, chewed-up teddy bear.

New people? They get the pink Easter Bunny, right after he shakes the stuffing out of it and slams it on the floor a few times.

But special people? People who might be bearing dog treats? They get the "Bear in the Bag." This is a mint-condition, black-and-white Panda bear Jake got for Christmas. Since he's mangled most of his other toys, I forbid him to take this one out of its bright red gift bag. The ribbon still holds the paper handles together at the top, so he can't actually get his teeth on the bear, just the bag. So the "Bear In the Bag" has become the highest honor bestowed on guests, once. No treat, and you can count on getting that stupid bunny the next time.

Although I give him hell for doing it, I love to watch him rush a farm fence and bark in frustration at all the cows. They're all out of place and slouching, but he can't get in there to get them organized. It frustrates him to a high-pitched whine. Chain link is hell on a herding dog.

And horses! Whoa, we do not like horsies. Jake could ride around Niagara-on-the-Lake for hours giving those carriage horses hell. Somewhere way back, horses must have taken the contract for the cattle drive away from herding dogs, and they've never been forgiven.

Jake may be a herder—at least he runs around and barks excitedly at all animals bigger than him—but it's certain he's no hunter. Out on walks along the railway tracks he runs with his nose to the ground like a crazed carnivore who's missed lunch. All told, he has walked within a yard of a still rabbit, a frightened kitten and a possum doing a poor job of playing dead. He has missed a fox chasing a groundhog across a frozen pond and chased a low-flying female robin who successfully lured him away from her chicks scattered on the ground.

And on Camelot Beach in an early-morning mist, he immediately sat down without being told and never moved a muscle without being leashed to watch two white-tailed deer frolic along the shallow shore for fifteen magic minutes. Once they were gone, disappearing up Sandhill, he took off chasing their fresh scent, but until then stayed silent and motionless, enjoying them as much as I did.

I love to watch him roll in fresh-cut grass. And even better, new-fallen snow. The look of sheer joy on his face as he travels around the yard on his back is enhanced by the grass clippings or snowflakes on his face. In the winter, he always looks like he just stuck his head in a bag of flour.

During a summer tournament at the Port Colborne Tennis Club, Jake nearly passed out from chasing balls and kids in the hot afternoon sun. So I walked him over to the shade of the arena, and he discovered a mound of icy heaven in the midst of the sweltering heat.

When he first came to the pile of snow dumped there by the Zamboni, he touched it tentatively, not believing his eyes. Then he turned and gave me that look like somebody really screwed up this time: snowdrifts in August !?! He tasted it, stuck his head straight in it and then rolled around on his back until the hill of snow was a puddle. I have a photo of him buried in this bank of snow, and the look on his face is pure rhapsody.

Jake loves running at people he hasn't yet met, seriously and with a menacing guise in his gait. Then at the last second, he jumps up and kisses them on the face. I know, it's not polite and I try to stop him, but he usually wins over one more friend for life. And this dog can cultivate friendships like no popular person I know.

The first time I asked my friend Dave Hurst if Jake could tag along for dinner, I know the doctor said yes just to be polite. Dave was lukewarm on dogs.

Now if I show up at Dave's without Jake, he makes me go home and get him. During dinner in front of the fireplace, Jake sits at attention between Dave's knees as a kind of substitute dinner guest. He's ready if Dave needs him.

The first casualty of having a dog in the house is privacy.

The shower once was a ten-minute, feel-good escape from the phone, the fax, the neighborhood kids working their way through school on my sponsorship money. Now the first minute of my shower is spent trying to convince Jake he's not going to join me and then, dripping wet and blindly reaching for towels, I somehow have to get him to move off the bath mat where he lies like a bodyguard.

On the john, in the morning, used to be a time to think or read. Now Jake bursts through the door, and there's no doubt about it—he's got me where I can't get away.

*Option #1*    He stands with his back to me, bumping backward into me saying: "Rub my bum."

*Option #2*    He backs up even closer to me, standing, waiting: "Scratch my back."

*Option #3*    He sits with his paws in the air in a begging position and leans his back against me: "Scratch my chest and belly."

This happened once, and I went along with the three-step program, merely out of curiosity. From day two it has been ingrained in our morning rituals as important and predictable as my mug of tea and his bowl of breakfast, my morning paper, and his yawning and stretching routine.

During those infrequent moments when he is not at full tilt chasing squirrels, waves, seagulls and other dogs, Jake is like a living alarm system. He sits perfectly still except for his ears, which become fully extended and erect as they turn quickly, simultaneously, a quarter turn at a time like radar panels absorbing any and all sounds that might be coming our way. Nothing in the noise department, the nose twitches like a rabbit in an herb garden. Nothing in the smell department, he gives the horizon one last optic scan before he circles three times, hits the ground with a thud and a groan, and goes to sleep with his chin atop his front paws.

Honestly, I think he's just covering his ass. Everything was in good order on my watch, he's saying. I was long gone off duty at the time of the attack, sir. I left everything in tip-top shape.

And I can actually see my dog think. If I give him an ambiguous order or use two familiar commands that seem to contradict each other, I can see a pained expression on his face as the wheels in his head turn slowly. "Stick!" means go get the stick, and "Bring!" means come to me with it. But if the words come out too close together, he stops dead and looks at me with the "Pick one, stupid" look. And there's that moment where I know he's confused but working through it in his brain, trying to decide which response will please me most. When the plan is clear, Jake is focused and everything is straight away: head, eyes and feet. But when "Stick!" follows "Bring!" too closely, the eyes move side to side, the head goes back and forth, and he paws at the stick instead of picking it up, awaiting clarification.

Once when Monica and I were walking through Mayville, New York, on the shoulder of the busy highway that splits the town, the word "cross" came up in casual conversation. In a flash Jake was on his way across the road with cars honking and braking to avoid him. "Cross!" first and foremost, is the order to cross a street, whether it's in or out of his context.

When Monica stays over and cannot tolerate my snoring, she sleeps on the couch in the TV room. Ever the vigilant protector, Jake situates himself in the hallway in such a way as to watch both of us at the same time. You'd think his eyes would cross watching everything at once. Within an hour he's got half her couch, and now she has to deal with stereo snoring. But when he signed off duty, all was safe and quiet from the hall security post.

I never saw this happen, but two neighbors told me that while I was away one week last summer, my house sitter had a three-year-old out for a day at the beach, and he was playing with a ball on the breakwall. They were bouncing a ball along the concrete wall that drops 10 feet down to the beach, with no railing. The neighbors said Jake ran behind him only once while the kid got the ball and brought it back to the group. Every time after that, Jake ran up and down between the boy and the edge of the breakwall.

Nobody in the group instructed him or even noticed, but the neighbors were amazed—Jake running along the ledge, easing the kid to the center where it was safe.

Jake is smart and usually predictable except on those occasions when curiosity gets the best of him, and he does something stupid and unusual. Like the day we walked by the house of Samson, a dog of mythical size and great strength, who lived just down Lakeshore Road. Samson is a huge rottweiler who was chained to the front of his doghouse across from the Rathfon Inn, which Jake and I walk by each day.

As we walked by, Samson would give Jake a low, gravelly bark that sounded more like it came from a member of the bovine family. Jake, as he does with all dogs he passes, breaks into an excited trot and whines, like "I'd really like to stop and chat, but as you can see I'm not on a leash! I'm free, I tell you, free! Plus I have a life and a place to go, you big doofus."

Once the routine was set, I didn't pay much attention to Samson except that day when I heard him stop in mid-bark, and I looked over and saw him standing with his mouth open as Jake walked behind him and into his doghouse. Clearly confused, Samson didn't move except to look over at me, but I was as dumbstruck as he was.

Samson was unnerved and embarrassed, but he never moved, not toward his house or Jake who was still inside checking out the digs.

Not until Jake came out and left the way he'd come, from around behind the doghouse, did Samson start ranting and raving about how if Jake ever tried that again he'd beat the Border collie out of him. And Jake carried on as if nothing had happened, as if he was merely seeing for himself how the other half lived.

Samson was never quite the same when we'd walk by. His nasty and dangerous reputation had been blown in one quick home invasion. He was just a big misnamed teddy bear. He moved away shortly after that.

And I can't get enough of the "bring thing" Jake does.

Up north on an island near MacTier with my buddies Dick and Jeff Reuter, all of us huddling under a dripping umbrella and hovering over a sputtering barbecue, nobody would walk back up to the cottage to get a round of beer. On a cold, miserable night at the cottage, nobody was willing to step away from the heat of the grill and into the rain to retrieve what we acknowledged we needed most—more beer.

And through the screen door, without opening it, here comes my trusty dog. Jake comes gingerly trotting down the long steps and into the bush with a Loblaw's bag in his mouth that contains three cold bottles of lager. That's his best trick of all—"Beer in the Bag!"

I don't care how many goals your kid scored last season, you cannot impress another man more than with a dog that brings you beer. That's my boy! (An assist on the play goes to Monica.)

The next day, the fun ended when my friends Dick and Bonna's genteel golden retriever, Keady, was bitten by a rattlesnake, although none of us knew it at the time.

Keady, who adores Jake, will run and swim beside him all day, trying to win his affection. Suddenly she shut down. Like a moping rag doll, she flopped onto the deck, lethargic and glassy-eyed. Jake, best in the role of rascal who tolerated Keady's flirting but did not return the romancing, immediately and uncharacteristically lay down beside her. And he never moved for hours, not until Keady was bundled up and whisked off the island.

She's fine today, but then nearer to death than any of us wanted to admit, Jake's spontaneous show of emotional support for Keady bared his soul, a deep dog spirit uncomplicated by human thinking. Like the Laings' two Border collies, dogs sense the time of our greatest need, and they comfort us. They fortify us. They know.

It's those quirky canine kinks, the lovable little foibles that surprise us and sustain us and make our dogs the greatest things in our lives.

They're all unique, of course, and you can immediately tell the difference between my dog and yours because my dog is the smartest, handsomest dog in the world … is too … is too. And yours comes a very close second.

# *The Dog Dictionary*

**PROBABLY, IN THE PAST,** I too have been insensitive to the canine cause, carelessly throwing around dog clichés like smoked-bacon treats during graduation ceremonies at the obedience academy.

However, now that I've become a dog fool … excuse me one second …"What's that, Jake? You want to have my supper instead of yours? Okay, but you'll have to drink my Murphy's Irish Stout out of your plastic dish. No glassware on the floor, big guy. That's the rule."

Where was I? Oh yeah, the meaning of phrases involving the word "dog" change drastically after you have one.

It used to be that if I worked hard all day, I'd be "dog tired" by bedtime. Now, after I work hard all day, walk Jake for one hour, throw the stick for a half hour, wrestle on the living room floor for fifteen minutes, I'm completely exhausted by bedtime. But I'm never dog tired, if you know what I mean.

This occurred to me one morning as I lay half-asleep, my face being warmed by Jake's panting. It's the gentlest alarm clock I've ever woken up to. Plus, it comes with a snooze option. I have five minutes between the time I roll over and Jake comes around to the other side of the bed and the wet kissing begins.

At that point I realized that any person who calls another person "dog breath" has never basked in one.

With the English language evolving to accommodate new words, and words changing meaning by the minute, I think it's high time we took a fresh look at dog diction.

- **dog's breakfast** ... the most important meal of a dog's day ... except for supper
- **dogbane** ... either a medicinal root plant or "dog bone" spelled incorrectly
- **dog cheap** ... nothing I've ever seen at the pet shop
- **dog-eared** ... the two furry appendages that stick straight up when you say: "Jawannagoferawalk?"
- **dog-eat-dog world** ... a dog barbecuing a wiener
- **dog face** ... a place where love resides
- **dogged** ... a word that best describes their work ethic
- **doggerel** ... the sound made by a dog with a tennis ball in his mouth
- **dog fish, dog tired** ... dog lie down for a nap
- **doggie style** ... nothing to do with fashion
- **dog tags** ... jewelry worn in American war movies
- **dogma** ... the dog's ma, often called a "bitch," nothing to do with demeanor
- **dogmatic** ... a term that describes how the doors open when a dog enters a supermarket
- **dogmatize** ... the process of training human beings to become dog fools

- **dog in the manger** ... the much-overlooked member of the famous Nativity scene who led everybody there in the first place but didn't even get the table scraps from the Last Supper
- **dog collar** ... an arrest made by a police dog
- **doggone** ... be back in a minute, the guy next door has food
- **dog kennel** ... a frat house for the four-legged
- **dog Latin** ... the canine rebuttal to pig Latin, as in *veni, vidi, I piddled.*
- **dog watch** ... the security dog of a dyslexic owner
- **dog wish** ... "A strange and involved compulsion to be as happy and carefree as a dog"—James Thurber
- **dog show** ... but dogs don't tell. Don't bring your dog to school.
- **dog trot** ... an early ballroom dance later made famous by fox
- **dogs of war** ... like the dog in the opening scene of *Gladiator* who got no credit and zero character development
- **dogwood** ... the best, softest stick in the kindling pile
- **going to the dogs** ... only about forty percent of all the food in doggie bags brought home from restaurants
- **hangdog look** ... it made Elvis millions (or was that the hound-dog look?)
- **a hard dog to keep on the porch** ... Bill Clinton
- **a dirty dog** ... formerly a cad and now a candidate for a shampoo and blow dry.
- **a hot dog** ... Oscar Mayer
- **underdog** ... where fleas go to get out of the sun
- **to rain cats and dogs** ... formerly the stupidest dog expression of them all, now pretty much the same.

- **let sleeping dogs lie** ... a misnomer since dogs, honest as the day is long, do not lie even when they're awake
- **the tail wagging the dog** ... a really, really good shake
- **top dog** ... the canine most commonly called "yours"
- **the triple dog dare** ... surpassing the single and outdoing the double, this is the defiant stare that declares dog war. You need a truce, Judge Judy and a good dog attorney

And finally, to lead a dog's life ... shorter but happier than a human life, living the life of royalty really without being dogged by scandals and paparazzi.

THEY BELIEVE THAT PETS

HAVE NO CONCEPT OF DYING,

AND THAT'S A BLESSING.

BUT THAT AWFUL CERTAINTY LURKS LIKE A

DARK CLOUD IN THE BACK OF THE MINDS

OF ALL TRUE PET LOVERS.

AND THAT IS OUR ONLY CURSE.

WWW.WILLIAMTHOMAS.CA